Race
An Introduction

Taking a comparative approach, this textbook is a concise introduction to race. Illustrated with detailed examples from around the world, it is organised into two parts. Part I explores the historical changes in ideas about race from the ancient world to the present day, in different corners of the globe. Part II outlines the ways in which racial difference and inequality are perceived and enacted in selected regions of the world. Numerous case studies, photos, figures and tables help students to appreciate the different meaning of race in varied contexts. End-of-chapter research tasks provide further support for student learning.

Suitable for courses focusing on the study of race and ethnicity, this textbook leads students through the formal structures, historical movements and everyday manifestations of race.

PETER WADE is Professor of Social Anthropology at the University of Manchester.

Race
An Introduction

PETER WADE

CAMBRIDGE
UNIVERSITY PRESS

CAMBRIDGE
UNIVERSITY PRESS

University Printing House, Cambridge CB2 8BS, United Kingdom

Cambridge University Press is part of the University of Cambridge.

It furthers the University's mission by disseminating knowledge in the pursuit of education, learning and research at the highest international levels of excellence.

www.cambridge.org
Information on this title: www.cambridge.org/ 9781107652286

First published 2015

Printed in the United Kingdom by TJ International Ltd. Padstow Cornwall

A catalogue record for this publication is available from the British Library

Library of Congress Cataloguing in Publication data
Wade, Peter, 1957–
Race : an introduction / Peter Wade.
   pages   cm
ISBN 978-1-107-03411-2 (hardback)
1. Race.   2. Genetics.   3. Sociobiology.   I. Title.
GN269.W22   2015
305.8–dc23

                                                                    2014050333

ISBN 978-1-107-03411-2 Hardback
ISBN 978-1-107-65228-6 Paperback

CONTENTS

FIGURES

TABLES

PREFACE

Despite the many years I have spent pondering and reading about the subject of race, I still feel a deeply rooted uncertainty about exactly what 'it' is and how to approach 'it'. This, I think, is no bad thing and guards against the oversimplification of something that, rather than being a single object, is a mercurial, shape-shifting and slippery set of ideas and associated practices, which exist always in relation to other phenomena. On the one hand, I have long been dissatisfied with approaches that side-step an effective definition of race, such that any specificity is submerged into general ideas about inequality and difference. This seems to lose a grip on something that is characteristic of racial concepts, despite their diverse and changing manifestations, which makes them different from other forms of distinction. On the other hand, I have also been uncomfortable with approaches that are overly specific and define race in terms of 'biology', 'colour' or even broader terms such as 'naturalisation'. These approaches tend to take for granted what all these terms mean and how their meanings change over time and space. My own reaction to this dilemma is outlined in the first chapter, in which I seek to pin race down, but in a way that acknowledges the precariousness and impermanence of the exercise, given that race is always on the move, is always embedded in a specific historical context and always exists in relation to other social connections and categories. These features undo an attempt to characterise race as a coherent object and point towards the existence of multiple processes of racialisation, in which very diverse social phenomena are freighted with racial meanings, which are still nevertheless recognisable as *racial* meanings, with their particular historical baggages and entailments.

My aim in this book is to place the concept of race in a very broad historical and geographical context, with a view to showing its affinities to other modes of social differentiation, which overlap with it, while also trying to grasp certain specificities and continuities. The limits on the length of the book have meant that the geographical range has not been as broad as I would have liked – more on race today in Asia and the Middle East would have been useful, for example – but it also seemed to me important to use detailed ethnographic data to show how categories and practices related to race actually work on the ground in the contexts that I do examine in detail. On the other hand, I felt it was important to spend time giving a broad historical overview, in order to grasp the very different things race has meant over the centuries, while also avoiding a narrative in which the late nineteenth and early twentieth centuries become the inevitable frame for thinking about what race is and what it does.

By training and by inclination I am an anthropologist, but I have purposely avoided putting the word anthropology in the title of the book, because race has to be addressed in a cross-disciplinary way. To deal only with the way anthropologists have engaged with the topic – from their participation in the science of racial (and racist) classification in the nineteenth century, through their pioneering role in dismantling this science and attempting to rigorously separate 'biology' from 'culture' in the mid-twentieth century, to their embarrassed avoidance of the topic after World War II, and finally their re-engagement with it in the last few decades – would be to produce a partial story. History is absolutely fundamental, to start with, while the insights from philosophy, sociology, geography, law and cultural studies have been central to the development of scholarship on the topic. In Part II of the book I have privileged ethnographic data, much – but not all – of it produced by anthropologists. This is because I want to give a grounded view of race in everyday practice, and anthropology specialises in ethnographic approaches. However, I also believe that anthropology is particularly well oriented to the relativising perspective I adopt here, which, by taking a broad historical and geographical view, questions concepts that may become taken for granted, such as nature, biology, colour and, of course, race itself.

ACKNOWLEDGEMENTS

I would like to thank Soumhya Venkatesan, Cecilia McCallum, Madeleine Reeves and Frank Dikötter for responding to queries about India, Amazonia, Russia and China respectively. I am grateful to the editorial team at Cambridge University Press for their support – Andrew Winnard, Helena Dowson, Valerie Appleby and Bethany Gaunt – and to the Cambridge University Press production team for shepherding the book into print. I would also like to thank the two anonymous reviewers who provided very useful feedback on the original book proposal and the reader who evaluated the final draft. This book was written during a period of research leave made possible by the freedom granted by a British Academy Wolfson Research Professorship, for which I am very grateful.

1 Knowing 'race'

I start by mapping out the basic conceptual territory of race, providing a guide to what kind of phenomena we, as anthropologists, and others are talking about when we use this term. This proves more complicated than we might have imagined, because race is a concept that has been used in varied ways at different times, in different places and by different people.

Before I start sketching out the territory associated with the concept, it is interesting to try a small experiment, to see what the term means to you.

DEFINING 'RACE'

This exercise is best done collectively, in a classroom or seminar context, but you can also do it individually.

Write down a definition of 'race', in terms of what you understand by the concept. What does something have to be, or to have, for you to apply the word 'race' or 'racial' to it?

What about the word 'racist'? Is this different and if so, how?

When I try this exercise with students in Britain, there is no consensus about these terms. Some people talk in terms of origins, nationality and cultural background; some mention religion; others mention physical appearance, referring, when pushed, to skin colour and perhaps type of hair. Occasionally, people will mention 'blood' or parentage. Being 'racist' is usually said to mean discriminating against someone – excluding them, insulting them – on the basis of these characteristics.

Already we can see that the term race covers a broad area, including terms – such as culture, nationality and religion – that might be seen as conceptually different from race. When we look at contexts where race is part of a public, official discourse and we might expect clarity about what it means, this looseness of meaning is reaffirmed. For example, Britain's Race Relations Act (1976) says race in its name, but actually defines 'racial grounds' of discrimination as including 'colour, race, nationality or ethnic or national origins'. Legislators were not trying to produce a coherent conceptual definition of race for the analytic purposes of social science, but their definition is indicative of the vagueness of the term. It also indicates a tendency to simply avoid a clear definition: 'racial grounds' includes 'race', which remains undefined, as if everyone already knows what it is.

In the United States, the Bureau of the Census regularly counts people on the basis of race. Answering the question 'What is race?', their website says

> The racial categories included in the census questionnaire generally reflect a social definition of race recognized in this country and not an attempt to define race biologically, anthropologically, or genetically. In addition, it is recognized that the categories of the race item include racial and national origin or sociocultural groups. (US Bureau of the Census 2013c)

Regulations in the United States specify at least five racial categories: White, Black or African American; American Indian or Alaska Native; Asian; and Native Hawaiian or Other Pacific Islander. In this case, then, 'race' refers to a specific set of categories, which themselves apparently refer to colour, ancestral origin and current geographical (regional, national) location. The categories are already 'socially recognized', but the Bureau also defines them in terms of 'origins': a White is 'a person having origins in any of the original peoples of Europe, the Middle East, or North Africa', while a Black or African American is 'a person having origins in any of the Black racial groups of Africa'. Ancestral origin is clearly the main criterion here (although, as in the British case, tautology or circularity slips in – race refers to origins in a 'racial group', at least for Blacks). It is worth noting that the categories have also varied over time: for example, in 1930 the main categories were White, Negro, Mexican, American Indian, Chinese, Japanese, Filipino, Hindu and Korean.

Brazil is one of the few Latin American countries to officially use the term race in its governance procedures: the census has had a question on *cor* (colour) since 1872, with occasional exceptions. The word *raça* (race) made an irregular appearance in the census over this period, but has been present again since 1991 in a question that asks people to identify their 'colour or race'. As well, recent legislation on affirmative action in favour of 'black' people (*negros*) makes reference to 'racial quotas' for university places and in 2010 a Statute on Racial Equality was passed. In this context race is generally understood as synonymous with colour, and the latter refers to a specific set of categories – white, brown, black and yellow – to which people are asked to assign themselves. Here, skin colour is the main criterion for defining race and, not surprisingly given the infinite variety of skin tones, no attempt is made to define these categories – they are assumed to be, to use the words of the US Bureau of the Census, 'socially recognized', even if there is not a social consensus on where the boundary lies between, say, brown (*pardo*) and black (*preto*).

If 'race' remains frustratingly vague and taken for granted, the term 'racist', by extension, is also hard to pin down. Few people today will admit to being a racist, yet accusations of racism abound. But what kind of discrimination do such accusations refer to? Apparently that based on race – but as we have seen for Britain, 'racial grounds' embraces colour, race, nationality or ethnic and national origins, a formulation reproduced, with the additional criterion of 'descent', in Australian and Hong Kong laws. Beyond the remit of the law, anti-Muslim attitudes in Britain are frequently branded as 'racist', even though strictly speaking they are based on religious criteria. The US Civil Rights Act of 1964 bans discrimination on the basis of 'race, color, religion, or national origin', but does not define race or specify how it is different from colour; presumably race refers to membership of one of the racial categories that are 'socially recognized' in the United States.

Brazil's Afonso Arinos Act (1951) banned certain types of discrimination based on 'race or colour prejudice' and its 1988 constitution outlaws discriminatory practices by reason of origin, race or colour, among other things. In Bolivia's 2010 Law Against Racism and All Forms of Discrimination, racial discrimination is defined as discrimination on the basis of 'race, colour, ancestry or national or ethnic origin', while racism is defined as a 'theory' that values 'biological and/or cultural differences, real or imagined', so as to benefit one group over another.

1.1 Chronology of race

One way of getting a grip, at least initially, on this ambiguity is to recognise that the concept of race has greatly changed over time and does not have a single meaning. I will explore this in later chapters, but it helps to give a brief timeline at the outset. Scholars disagree on the details of this, but I will outline a fairly standard chronology for the moment – although I will be taking a critical approach to this narrative later on.

The standard history traced by social scientists and historians for the idea of race, in its Western or Euro-American context, has three broad periods. In the first, between about the fourteenth and eighteenth centuries, race emerged in embryonic form and gradually developed. During this time the concept of race depended on a mixture of ideas about human nature, environmental influences and culture. Some scholars maintain that the concept of race does not describe thinking about human diversity during much of this period; they argue that the emphasis was more on culture shaped by environment than anything else.

There then followed a second period between about 1800 and the 1940s, when race was consolidated as a central pillar of Western thought and, above all, science. Biology – which became a recognised discipline in the early 1800s – was fundamental to defining racial difference, seen as physical differences that accounted for cultural diversity and moral qualities. For many scholars this is the classical period of the concept of race, when so-called scientific racism became dominant and the discourse of race was clearly about an underlying human biology, which defined a small number of major races and many smaller sub-racial categories. The races were placed in a hierarchy, with whites or 'Caucasians' at the top, their dominant position justified in terms of an allegedly superior biology. For many scholars this period really defines what race is all about.

Then, from roughly the 1920s, challenges emerged to this racial science, eventually discrediting it. A lot of scientific evidence accumulated over time to indicate that (a) human capacities, such as intelligence, were not linked to biological race; and (b) that human biological diversity could not be classified into the entities that had previously been called 'races'. Social scientists therefore concluded that race was not a biological reality, but instead a social category that used a language of biology and physical appearance as criteria for marking differences – race was a 'social construction' (see Chapter 4).

As a corollary of this shift, social scientists have observed that, after World War II (WWII), race has become masked and silenced in comparison to earlier periods. They note that, in many contexts, use of the term race seemed to evoke Nazi ideologies of white supremacy and other politically controversial racial ideologies, often associated with European colonialism and/or with the stark forms of legalised racial segregation and discrimination that operated in the United States until the 1960s and South Africa, under apartheid, until the 1990s. The term race therefore began to drop out of the public, political vocabulary, becoming an almost taboo word in some places (e.g. France, Germany), its very use smacking of racism. It was often replaced by the term ethnicity, understood to imply only cultural difference, or by some reference to 'cultural minorities'. As I noted above, students would often respond with the term 'culture' when I asked about 'race'.

Race might continue to be part of official discourse in some countries – such as Britain, the United States or Brazil – but even then it is often merged with ideas about ethnic (read cultural) and national differences. In South Africa, for example, after the fall of apartheid, Black Economic Empowerment policies continued to target 'black' people as a category, while the state statistics office decided that 'race' should be replaced by 'population group' as the preferred census term, defined as a group with 'common characteristics (in terms of descent and history)' (Statistics South Africa 2004: 12): here history is as important as biology in defining the group. Also, scholars observed that, even in the United States, where the term race remained current and was defined in the census in terms of origins, many people – especially but not only whites – adopted a 'race-evasive' discourse or acted 'colour-blind', trying to avoid talking about racial differences at all, as if these did not matter because 'we are all the same' (Bonilla-Silva 2003; Frankenberg 1993). Talking overtly about race and colour seemed to carry the danger that people might see you as racist.

Still, racism remained in so far as the categories of people who previously would have been, or still were, identified as racial groups continued to suffer discrimination: the 'population groups' of post-apartheid South Africa were the same as the 'races' defined by the apartheid-era Office for Race Classification. Social scientists therefore began to talk in terms of 'cultural racism', 'neo-racism' and 'new racism', in which an explicit discourse of race was absent, biology and even colour were not mentioned, and the talk was instead in terms of cultural difference (Barker 1981; Goldberg 2008: 216; Hale 2006: 144; Lentin and Titley 2011; Taguieff 1990; Winant 2002).

1.2 Is race defined by appearance, biology and nature?

Given that the meaning of race has changed so much over time, but faced still with the perceived need to define what race is as an object of study, social scientists have tried to give it a more specific meaning. They see race as one particular way of classifying people, among other ways. The idea is that people classify people on the basis of perceived differences of many kinds and divide them up in lots of ways – what gender they are, how they behave (often called 'culture'), how wealthy they are, where they come from, where they live, how old they are, what they look like, what gods or spirits they recognise and worship, what football team they support, and so on. Frequently, classifications are deployed to include certain individuals or groups and exclude others. Such exclusions may not have important material consequences – if you're not in one group of football supporters, you can likely be in another – but often they are part of hierarchical power relations of domination and subordination that shape people's access to resources, their freedoms and their life chances. It therefore becomes particularly important to understand the ways in which people classify people and what the consequences are.

Racial modes of classification are often said by social scientists to focus on 'biological', 'physical' or sometimes 'phenotypical' differences (phenotype is the entire physical organism, but often refers to physical appearance). One anthropologist (Fluehr-Lobban 2005: 20) says that 'race is about outward physical appearance or phenotype'. Sociologists Anthias and Yuval-Davis (1992: 2) contend that racial differences are those constructed on 'the basis of an immutable biological or physiognomic difference ... grounded on the separation of human populations by some notion of stock or collective heredity of traits'. Goldberg states that race 'has always had to do ... with the set of views, dispositions and predilections concerning culture, or more accurately of culture tied to colour, of being to body, of "blood" to behaviour' (2008: 175). The emphasis is on colour, bodies and 'blood', and Goldberg highlights that race links these physical traits to culture, being and behaviour, which are thus made to seem natural.

These definitions in terms of physical appearance and biology are very common, but not everyone is happy with this emphasis. In an earlier formulation Goldberg took a slightly different line, seeing talk only of biology and appearance as too narrow. He disagreed that 'ideas about race are inherently committed to claims about biological inheritance'. Instead, race itself 'does not concern biological but naturalised group relations' (1993: 72, 81): a racial classification is one that sees differences between people or groups as 'natural' in some way, without these necessarily being understood as biological (for example, they may be seen as God-given, determined by the environment or 'the stars', or simply seen as 'the way of the world'). The focus on naturalisation is broader than that on biology.

Shanklin also takes a more open approach, defining racism as a kind of prejudice 'directed against those who are thought to possess biologically or *socially* inherited characteristics that set them apart' (1994: 105, emphasis added). Going further still, Hartigan simply says that race is 'a system of classifying people into groups, either explicitly or implicitly promoting the notion that these groups are ranked in terms of superiority or inferiority' (2010: 211). Hartigan avoids specifying what a racial classification is, as distinct from any other hierarchical classification.

These moves away from a focus on biology clearly reflect changes that social scientists have detected in the concept of race since WWII, as outlined above. If we are now in an era of 'cultural racism', when references to biology tend to be more hidden and when even overt reference to aspects of physical appearance, such as colour, may be evaded because people fear it smacks of racism, then we cannot limit our definition of race to biological criteria. We still intuitively want to include certain phenomena in the field of race and racism, even though there is no overt mention of physical appearance or biology. This is what lies behind Goldberg's focus on naturalisation. In cultural racism differences of culture could be naturalised, as if they were an almost innate and essential part of the person or group: tastes for certain foods and their smells, particular forms of family life and values, religious beliefs and practices could all be associated with specific categories of people in this naturalising and 'essentialist' way (a way that made these characteristics seem an essential part of a person or a group; see Chapter 5 for more detail). Gilroy argued that, in Britain, 'culture [is] almost biologised by its proximity to "race"' (1987: 61). When culture is naturalised in this way, cultural racism contends that it is 'only natural' for people to want to live with others who are culturally like them.

1.3 Culture, appearance and biology revisited

These ideas about biology and culture as the defining features of race are very helpful, but they suffer from a number of problems. First, if we focus on culture, social characteristics and superiority/inferiority, trying to connect with the tendencies of post-WWII 'cultural racism', then we begin to lose a grasp of what makes race different from other classifications and rankings, which may take nationality, culture, wealth, education or other such criteria as their basis. Surely we need some kind of specificity here? How can we tell if we are confronting *racism* here, rather than nationalism, xenophobia (fear of the foreign) or ethnocentrism (belief in the superiority of one's own ethnic group or culture)? Many scholars do not address this question directly, instead assuming that everyone will know what counts as 'racial'. But, implicit or explicit, the answer is that (a) cultural racism is directed at groups and individuals who previously would have been subject to a more explicitly racial discourse (in terms of biology, appearance, etc.); and (b) cultural racism depends on naturalisation and/or tacit references to physical appearance or biology. We are brought back to these defining criteria.

Second, then, if we do plump for appearance, biology and nature as the defining features of race, we immediately encounter the problem that all these are terms that include a host of possible human differences, most notably gender differences. Ideas about the differences between men and women have generally made much of actual and imagined differences in appearance, biology and underlying nature. Ideas about the difference between young and old people also often depend on notions of biology and nature. But clearly we do not want to confuse race with gender and age.

Linked to this is the fact that plenty of differences in physical appearance among people are not understood as relevant to 'race'. It is specific aspects of physical appearance – typically skin colour, hair type and certain features of the face and head – that are understood by social scientists to indicate when a 'racial' classification is at work. Other aspects, such as height, wrinkliness, double-jointedness, length of fingers, fatness, thinness and so on, are rarely seen to have meaning as 'racial' traits.

We can make explicit what is usually kept implicit in the definitions cited above and say that, in general terms, the physical differences that have generally become part of racial classifications are the ones associated with the geographical diversity of humans that has emerged through evolutionary history as they have spread across the globe. People have adapted to their environment in ways that are reflected in aspects of phenotype, which also tend to be passed on in hereditary fashion, such that human phenotypical variation has, over time, come to have a very broad association with geographical environment – although that association is complex and variable, not neatly parcelled into clear categories. This type of phenotypical difference still includes a huge variety of possible traits, but certain ones may be selected out by people as significant markers to make distinctions. I say 'may be' because, as we will see later on, the differences that are made to count, by those doing the classifying, vary a great deal according to context. Physical differences that might seem important given the course of history, such as skin colour, have not always been seen as very significant. The important point here is that, in order to understand how appearance figures in racial thinking, we have to understand specific histories, which are generally histories of human encounters and, frequently, ones involving power, domination, subordination and conquest. Appearance is not a simple objective defining feature that can be taken for granted. A definition of race as ideas that refer to physical appearance raises as many questions as it answers: which physical differences are perceived as significant, why and with what effects? Those questions have to be addressed by looking at particular historical contexts.

The question of appearance indicates a third issue, which is that terms such as nature, heredity and biology are highly complex concepts, with varied meanings. If appearance turns out to be more complicated than it seemed at first sight, 'nature' is even more so: anthropological and historical studies show that what counts as 'natural' in humans and what is implied by something being 'natural' in humans has varied over time and place according to people's concepts of how the world works. For example, in the West nowadays we tend to operate with a view of human nature as a kind of underlying biological reality, on top of which is plastered all the 'culture' we acquire as we grow up. In previous eras – for example, in the eighteenth century and before – this distinction was a lot less clear; things that we now might see as cultural (e.g. a person's moral qualities) might be seen as part of his or her physical constitution, which could also be passed on 'in the blood'. If race is defined by its reference to 'nature', then recognising when nature is being invoked and understanding its significance involves grasping what nature means in a given time and place.

The same argument applies to the term heredity. We have seen that appearance and heredity are closely linked, in that the aspects of appearance that become racial markers are usually ones that, because of hereditary transmission between generations, have some continuity over time at a group level (which is not to say that all hereditary traits of appearance become racial markers). Heredity is key to racial thinking, because it provides a way of thinking about connections between some internal essence (such as 'blood'), outward appearance and behaviour. Each member of a perceived group or category has some of this essence and passes it on to offspring through sexual reproduction. The essence is usually thought to express itself in physical appearance (although it may be seen as hidden 'inside' and thus invisible) and in behaviour. But, like nature, human heredity is a concept that varies a lot according to historical and cultural context. Today we understand this in terms of the transmission of genes. Although parental genes get mixed up in the process of sexual reproduction, the genes themselves do not change. Pre-genetic concepts of heredity in the West generally thought that people (and animals) could pass on to their offspring traits that they had acquired during their lifetime, making heredity a much more malleable process than genetics decrees.

A similar point can be made about biology, especially as the word did not exist until the early 1800s; if we are talking about race before that time, we will not find explicit references to biology as such. The very term biology indicates a way of thinking about humanity (and the world) in which the natural, biological realm can be clearly separated from everything else (culture, morality, gods).

For all three terms, nature, heredity and biology, there is a common tendency to assume they connote permanence and fixity in relation to humans. If something is 'in the blood', natural or biological, it is said to be fixed and hard to change; it is 'human nature'. Many definitions of race note that grounding human differences in biology or nature lends them a fixity: 'race gives to social relations a veneer of fixedness' (Goldberg 1993: 81); the attribution of social meanings to physical variations is based on 'a notion of heredity and permanence' (Smedley 1998: 693). But if ideas about physical variation and nature vary according to time and place, it may also be that the fixity we attribute to biology is not necessarily present in other contexts.

Related to this, we often make restrictive assumptions about what constitutes a 'physical difference' in appearance, limiting this to fixed phenotypical differences. At other times and in other places important differences in appearance might include hairstyles, body decorations and modifications (e.g. piercings) and also clothes. These are things that we in the West today might see as simply 'cultural', but in other contexts might be seen as a more inherent or 'natural' part of the person or group and that might be deployed in classifications that seem to be 'racial'. Overall, it is important not to take terms such as physical appearance, nature and biology for granted, even if we do want to use them in our definition of race.

1.4 Race, comparatively and historically

We have seen that the way the term race is used today – for example, in official policy and legislation – is very loose: it often goes undefined, as if everyone knew what it meant, and it is often deployed alongside related terms such as ethnicity, origins, nationality, culture, with little clue as to what all these terms mean and how they relate to each other. Social scientists try to inject some rigour into this, by saying that race is a way of classifying people that generally uses perceived or imagined differences of physical appearance, biology or human nature as criteria for the categorisation. Culture, behaviour and modes of being are classified too, but the defining basis for the classification is 'colour, bodies and "blood"'. This is usually said to have the effect of making the classification seem natural and durable.

This is a big step forward, but we need to be careful about assuming that *any* reference to biology, nature or physical appearance automatically signals 'race'. If it is only *some* such references that signal 'race' then we need to know why these ones, and how they come to play this role when other biological criteria signal other things (gender, age, etc.). We also need to be careful not to assume that we automatically know what is at stake when a given mode of classification refers to nature, appearance or biology: fixedness may well be an effect, but we need to be alive to other effects too ('nature' might be malleable, such that a person's 'race' could change or could alter its meanings).

A further difficulty emerges when social scientists are faced with 'cultural racism' because then a phenomenon that we want intuitively to include in the domain of race and racism seems to lose – or at least hide from clear view – the key features that are often said to define race, that is, biology, heredity, physical appearance and naturalisation. Why should we continue to talk about race and racism in these circumstances? Is it because the categories of people suffering or practising cultural racism are often the same as before: blacks, whites, Asians, Native Americans, Aborigines, coloured people, etc.? Or is it because the cultural differences at issue become naturalised and essentialised? Both these questions suggest relevant answers, but it is not a simple matter.

The best way to approach this is (a) comparatively and (b) historically. The first step is to place racial classifications – understood broadly, without a restrictive definition at this point – alongside other ways in which people classify or have classified people, which seem to share some characteristics with what we think of as race, such as an emphasis on physical appearance or bodies. This poses the question of how widespread racial thinking is and what we want to include in the category 'race'. The second and related step is to ask if we want to see race as having a specific history that is rooted in 'the West', whether that begins with the appearance of the word in various European languages in the thirteenth and fourteenth centuries, or rather begins with European colonialism and the Atlantic slave trade in the sixteenth century, or perhaps only properly emerges as late as the nineteenth century when European science enshrined race as a key explanatory concept for understanding human difference.

1.5 Comparisons

Anthropological and historical studies allow us to look at a number of ways in which humans classify humans, in ways that seem to share some of the features that, with qualifications, we have noted as relevant to race: physical appearance, heredity, nature, biology, essence – all linked to culture and behaviour. We will see that there are examples of classificatory practices that essentialise, naturalise and locate difference in or on bodies. The question is whether we want to include these as examples of racial classification; is it analytically useful to do so? The following examples look at caste in India, *zu* and *zhong* in China and alterity (otherness) in indigenous Amazon societies. These – especially the first and the last – are all contexts with an abundant literature, which I draw on very selectively to make the comparative argument that concerns us here.

Caste in India

Caste has proved an interesting place to make comparisons with race. Simple comparisons are complicated by the fact that caste, like race, is not a single thing or abstract system, but instead a very varied set of ideas and practices that have changed over time in South Asia (and in the South Asian diaspora). The attraction of caste as a point of comparison lies in the fact that, as a practice of classification, it involves dividing up people into ranked groups, which, at least in principle, share bodily substances that are seen as heritable and that define certain key aspects of the person, for example in terms of 'purity' and 'impurity'. A person may be defiled by touching people of lower caste status, sharing food with them, being touched by their saliva, or even their shadow. Accordingly, castes are supposed to be mainly endogamous, although in practice all sorts of accommodations may be made. Castes are also linked to occupations in a society-wide division of labour, with certain tasks seen as low caste (e.g. service, labouring, waste removal) and others seen as high caste (e.g. priesthood, scholarship, the law).

Bodily substance may be conceived of in terms of 'blood', which can suggest ideas of biological heredity and essentialism to outside observers. But it is important to grasp that bodies are thought of in terms of multiple substances (saliva, bile, phlegm, maleness, femaleness). In any case, bodily substances and their purity are influenced by food, contact, behaviour and interaction with innumerable other substances in the people and things around one, all of which have to be carefully regulated to maintain balance and minimise defilements, which have to be corrected through ritual purifications (Barnett 1976; Daniel 1987). Thus bodily substance is quite a malleable thing: because something is 'in the blood' or 'in the body' does not make it fixed. On the contrary, bodies and their substances have to be continuously regulated and cared for.

Physical appearance does not play a major role in defining caste status. There are thousands of *jatis* or castes in India, so physical appearance can hardly be a clear criterion. However, lightness of skin is socially valued and nowadays there is a big market in skin-lightening treatments in India. There is a broad correlation between status and darkness of skin. In ancient India colour was associated, probably symbolically, with the four major *varna* or original castes defined by the ancient Hindu scriptures: Brahmans (priests and scholars, coloured white); Kshatriya (kings, governors and soldiers, coloured red); Vaishyas (cattle herders, agriculturists, artisans, merchants, coloured brown); and Shudras (labourers and service providers, coloured black). There was also a category for foreigners or 'barbarians', *mlecchas*, but this does not seem to have been associated with a colour (Brockington 1995).

Skin colour aside, other aspects of caste do seem to resonate with race: categories ranked in terms of status and power, heritable substance or essence associated with certain behaviours and restrictions on marriage. Caste involves naturalisation, in the sense that the order is seen, at some level, as ordained by gods and is based on a concept of bodily heritable substance. Whether one wants to call that 'biology' or not is a moot point, as it depends on when such a concept entered Indian ways of thinking about the body, and especially on whether the assumption is that 'biology' connotes fixity.

A judgement as to whether caste is like race or not has to contend with the fact that European ideas of race had a major impact on India during the colonial period, and especially in the nineteenth century, as European scholars and administrators interpreted Indian social structures through the optic of their theories about human diversity, which at the time used race as a central and highly elaborated concept. In an important sense, British rule consolidated and rigidified the caste 'system' – with the help of certain groups of Indians who saw this as a useful development – while also injecting an explicit language of race. This language was then taken up by Indian nationalists and some religious fundamentalists, promoting the idea of a good 'Indian race' linked to Hinduism (Bayly 1995, 1999; Robb 1995b).

But according to historian Peter Robb, caste's 'race-like essentialism was not – or not only – a nineteenth-century borrowing' (1995a: 61). The central concern with essential, hierarchical, distinctive traits and roles that are inherited is 'race-like', when race is defined as 'any essentialising of groups of people which held them to display inherent, heritable, persistent or predictive characteristics, and which thus had a biological or quasi-biological basis' (Robb 1995a: 1). The essentialism involved in Indian reckonings of differences among people is not, however, always clearly 'race-like' in this sense, because, as outlined above, blood and behaviour are seen as linked closely together, such that blood shapes behaviour, but behaviour also shapes the (purity of) blood. Robb's emphasis on 'biology' introduces possible complexities, because it is not clear what counts as biological in the Indian context. Robb's addition of 'quasi-biological' recognises this problem and opens up the definition, but he does not offer any further detail as to what counts as 'quasi', so we are left in the dark.

An early commentator on race and caste, the sociologist Oliver Cox, saw the two as very different (Cox 1948). For him, race was linked to a capitalist market system, in which racism limited the opportunities of certain categories of people. This was against their will and against the fundamental tenets of freedom that were supposed to govern liberal, capitalist societies and defined racial exclusion as wrong. Caste, in contrast, involved a division of labour and marriage rules that were agreed and in tune with the overall norms of the society, which defined caste arrangements as right. In India a person was either a member of a caste or not; there were no 'half-caste' people; the natural substance of caste was not divisible. In race systems 'half-race' or mixed-race individuals were a common, and indeed inevitable, category.

In a critique of Cox, Williams (1995) argues that caste and race do have common elements as systems of classification that work to divide people up into categories and then direct resources to these categories or block such access. Both systems merge the opposition pure/impure with the opposition superior/inferior. Both systems imagine essential human substances, which everyone shares (e.g. blood), but which are differentiated to make people both similar (same caste, same purity, same race) and different (different castes, different purities, different races). These substances are passed on through sexual reproduction and shape people's behaviour, place and status.

In sum, caste has certain affinities with race, but is also very different. Before deciding how to proceed, let us examine China.

Zu and *zhong* in China

A similar set of issues arises in relation to China. There is little doubt that complex discourses of race existed in China (and Japan) from the late nineteenth century, which were shaped to some extent by Western ideas. But historian Frank Dikötter (1992) argues that the Chinese had discourses of race that preceded Western influences and therefore that

Western ideas fed, usually in a very selective way, into developing Chinese ideas about human diversity (on Japan, see also Takezawa 2005).

Dikötter thinks that 'phenotypical variations', 'biological differences' and 'physical features' are the key criteria for defining race. He uses the word race to translate a variety of Chinese terms – all of them related to the key terms *zu* (lineage, descent group, ethnic group, 'race') and *zhong* (breed, kind, seed, biological type, species, 'race'). These terms 'appear to stress the biological rather than the sociocultural aspects of different people' (1992: viii–ix).

When discussing ancient China, however, he says that the Western 'dichotomy between culture and race' (by which he means biology) should be abandoned, because 'physical composition and cultural disposition were confused in Chinese antiquity', which immediately makes it hard to see how he could tell which terms 'stress the biological' (1992: 3). Still, he shows that the ancient Chinese distinguished in very clear terms between themselves and 'barbarians' who did not follow Chinese ways. The discrimination was primarily about life-ways or 'culture', and in principle life-ways could be changed. But the understanding of life-ways was strongly naturalised. A fourth-century BC saying had it that 'If he is not of our race, he is sure to have a different mind' – the term used was *zulei*, a combination of *zu*, which evokes descent and genealogy, and *lei* (kind, type, class). Various alien groups were compared with animals, and both the Chinese and others were said, in the third century BC, to have their own 'nature, which cannot be moved or altered' (1992: 3–4). A fifth-century AD writer opposed the 'inborn nature of Chinese', which was pure and harmonious, to the 'hard and obstinate nature' of foreigners (1992: 19). A text from 1656 emphasised that 'Chinese and barbarians are born in different places, which brings about the differences in their atmospheres ... [and] their customs'; in different places, 'the ether is different, people have a different essence, nature produces different things' (1992: 27). This clearly shows the naturalisation of 'customs', linking them to environment and human essence.

The barbarians could be both white and black, but the Chinese saw themselves as white, so Westerners, who were as odd as Africans, were termed 'ash-white' as well as hairy. Blackness was associated with slavery, although slavery was not only associated with Africans, who seem to have been present in very small numbers as early as the tenth century AD.

Dikötter sees late eighteenth- and nineteenth-century ideas developing on the basis of these images of the barbarian. The Chinese elite developed a complex and rich series of stereotypes, now linked to a broader global geography and with more detailed (but not necessarily realistic) images of Westerners, South Asians and Africans. There was increasing contact with the West, but Dikötter argues that this had a relatively minor impact economically and intellectually. Europeans were generally seen as white, but as weird, physically bizarre and scary: hair (on the head, face and body), colour, size and other bodily features, alongside sexual behaviour and other customs, all received intensive attention. Chinese envoys to foreign parts wrote of their horror and fear at the bodies and behaviour of non-Chinese. The idea of the Chinese as a 'yellow race' was popularised in the West in the mid-1800s and it became increasingly common in China too, but not just as a Western import; yellow was seen as a good colour that already had strong positive connotations in China. By the late nineteenth and early twentieth centuries Chinese intellectuals were using terms such as *zhongzu* (combining the idea of lineage/descent with that of biological type) to talk about race in terms clearly influenced by Western race theory (learned by studying abroad or via a few translated texts), but Dikötter argues that the basis for this kind of thinking was already there and that the Chinese adopted Western ideas very selectively.

As with the Indian material, there are lots of reasons why we might want to say that Chinese concepts of *zu* and *zhong* are ideas of race, even before the influence of Western racial theories. But should we define these concepts as 'racial'?

Amazonian alterity

It may seem odd to go to the indigenous people of the Amazon as a comparative example for thinking about race, but I do so because of the importance they attach to classifying people and things as 'other' and also to bodily processes and substances. Many anthropologists have noted that Amazonian peoples attach great importance to 'alterity', the condition of being 'other' and opposed to the self or the in-group. Many things can count as other. For the Piaroa of Venezuela others include spirits, gods, animals and non-Piaroa; within the Piaroa others may also include shamans, and exchange partners and potential marriage partners outside the community. These others can be seen, in varying ways, as dangerous and powerful; distant others may be seen as violent, cannibal, sexually profligate, monstrous, morally wicked, excessive, and lacking in reason and proper behaviour (Overing 1996). Yet others are also very necessary for social life. The 'very Amazonian message' that Overing takes from the Piaroa is that 'the achievement of sociality requires those different than self'. Others have to be engaged and partly incorporated in order for human life to be social and productive.

> Without alterity there is no productive capacity. The Piaroa ... understand sexual relations as barren unless between people essentially different from one another. Thus both affines [people outside the community, who become related through marriage] and exchange partners are essential to the creation of human productivity. (Overing 1993: 205)

Overing does not tell us exactly what constitutes an 'essential' difference from a Piaroa perspective, but we learn that it must be something quite flexible and actually not very 'essentialist' in the sense of being fixed. People of the local community are classified as 'the collectivity of like beings to which I belong'; in-marrying affines are by definition 'alien' to that collectivity, but they 'become physically "of a kind" through the process of living together' (Overing 1993: 206).

For the Cashinhua of Brazil, people inside the local settlement are 'real kin'; the settlement and its gardens are the 'inside'; the 'outside' is where strangers, game animals, traders, spirits and male affines exist; the outside is marked by relations of seduction, male affinity, violence, repulsion, trickery, predation and difference (McCallum 2001: 70). But 'it would be unfair to accuse the Cashinahua or other indigenous Americans of ethnocentrism, for "otherness" is not seen as an essential attribute of bodies or as an inescapable physical condition' (2001: 66); otherness can be transformed through the work of being together, eating together and procreating together. For Amazonians bodies are indeed 'the main site of differentiation between different life forms', but bodies are composite forms, which are not born ready-made, but are constructed through the input of substances and affects provided by parents and kin, a process in which the incorporation of the other is vital (Santos-Granero 2009: 7; see also Santos-Granero 2012). For the Wari of the Brazilian Amazon a non-Wari woman can become Wari by bearing a Wari man's child, which transforms her blood: 'corporeal substances impart qualities of identity to those who incorporate them' (Conklin and Morgan 1996: 668).

Concluding comparisons

My purpose in discussing India, China and the Amazon has been to explore contexts in which people outside a Western context (although also in contact with and shaped by Western influences) think about and classify certain aspects of human variety. We can see that in all these contexts there is a concern with difference and similarity, thought of in terms of bodies, 'blood', behaviour, values and environments.

The Amazon serves as an example of a concern with all these things, but in a way that I do not think is 'race-like'. There is what a Western mode of thought might want to call 'naturalisation' in the sense that human qualities are linked to animals, spirits and places, but none of these are separated from the realm of humans, such that the former could be said to determine or even shape the latter. There is a different conception of both nature and culture, which does not lend itself to seeing one as an underlying reality that shapes the other. There is a powerful process of othering, with others often being classified in what look superficially like essentialist ways and also sometimes classed as inferior or threatening. But essentialism is undone by ideas about the way bodies and their essences are continuously made and made up; while the other is something to be engaged and incorporated productively. Overing (1996) contrasts Western and Amazonian modes of othering precisely in terms of the Western concern with hierarchy and domination, in which the other is inferior and should be conquered. Even when Western theories posit the 'other within', as psychoanalysis does (the childlike, the unconscious, the primitive mind hidden within the adult psyche), the other is located in a hierarchical order and placed in a position of inferiority. According to Overing this is not the case in Amazonia.

India and China are somewhat different. In these contexts, before Western theories of race had a significant impact, there is a clear sense that local ways of perceiving and classifying human difference included a certain 'race-like' essentialism, based on ideas about heritable, collective sets of traits, which characterised groups or categories and the individuals in them. These traits were seen as located in bodies (although not necessarily linked in a simple way to skin colour) and connected to 'blood', understood as something that connected generations through sexual reproduction, and often to environment. Much of this resonates with race, whether defined intuitively or by social scientists. But it is also important that bodies, blood and heredity are not seen as fixed, especially in

India, apparently less so in China (at least in Dikötter's account): they are all malleable. In ancient China, Dikötter notes that physical composition and cultural disposition were fused together, so that we cannot separate out the 'biological' from the 'cultural'. This might lead us to say that these ways of thinking are not truly 'racial'.

My view is that this is not the key feature, because, as we will see in later chapters, such an intertwining and blurring of what, in the recent thinking of the 'modern West', would be separated out as culture and nature (or specifically biology) is actually characteristic of Western thinking about race (and things more generally), especially before the nineteenth century and even, in more limited ways, during that century and after. Instead, I think the key issue is whether it is *analytically useful* to encompass these phenomena and contexts in the domain that we wish to call race studies. My view is that it is useful to be more restrictive and to narrow our focus onto a specific history that is associated with Western modernity (defined broadly), *as long as* we recognise that this is an *artificial* tactic, because modes of thought found in other times and places may share key elements with racial thinking.

This immediately raises the question of the history of race and how the concept became constituted in Western history. We have already broached historical issues, by exploring China and India before they were significantly influenced by Western ideas of race, but now we need to explore race in relation to a Western history.

1.6 Race in the history of Western modernity

The emergence of the concept of race will be explored in detail in the next chapter, so here I will only outline how these issues affect the way we define race and shape our sense of what the domain of race studies encompasses.

We are faced with the same problem that arose in the previous section on comparisons: for a 'Western' context, if there are historically specific ways of classifying people and thinking about human difference, where do we draw the lines between those that are non-racial, those that are race-like or proto-racial and those that are properly racial? My answer is roughly the same as for the previous section: we cannot draw hard and fast lines, because, whenever we locate its emergence, race has changed a great deal over time and some of its key features can be found in a variety of historical contexts that are often thought to precede the emergence of the concept of race. This concept drew on existing ways of thinking about human difference and developed them into an evolving mode of classification and practice.

Race, colonialism and Western modernity

If focusing on modes of categorisation does not delimit race in a specific way, a focus on particular histories does help us to sharpen the concept. The issues at stake here revolve around the idea of Western modernity. While we have seen that some historians of South and East Asia identify racial or 'race-like' modes of thought outside the compass of Europe and its history, many more scholars see race as a concept and racism as a practice as having emerged in the West, and specifically in conjunction with European colonialism and domination: the conquest of the Americas, the African slave trade, the abolition of slavery, imperialism in Africa and Asia, Anglo-Saxon expansionism across the United States, the global expansion of capitalism and so on (Eze 1997; Hall 1992; Hannaford 1996; Smedley 1993).

As we saw above, Oliver Cox (1948) linked race to capitalism and its exploitation of labour, including slave labour, for profit (see also Miles and Brown 2003). Another early theorist, Frantz Fanon, linked colonialism and racism, famously exploring the psychological impact of colonial racism on subordinated people (Fanon 1968, 1986 [1952]). Stuart Hall, inspired by Fanon, links the emergence of racial thought to European colonialism (Hall 1992); he also connects twentieth-century racism to the operation of cultural hegemony in capitalist societies, that is, to the way domination operates not only by direct force, but by indirect means (involving beliefs, values and culture), which make domination seem natural and common sense (Hall 1996a, 1996b).

A few scholars highlight more domestic, European roots for race and racism, focusing on late eighteenth- and nineteenth-century shifts in ideas about governing, which began to treat national populations as biological units, which also had to be protected against supposed enemies and degenerate elements within the nation, perceived threats that became the objects of racism (Foucault 1998, 2003; Taylor 2011). Whether race is linked to external colonialism or internal threats, modernity is the main frame of reference, albeit a shifting one: some see modernity as starting in the fifteenth century, with the 'discoveries' by Europeans of Africa and the Americas, while others emphasise subsequent periods, such as the eighteenth-century Enlightenment period of scientific discovery (see Chapter 3), or nineteenth-century imperialism and industrialism.

When linking race and modernity, an important insight sees them as necessarily connected, rather than racism being an aberration or an unfortunate contradiction in relation to the ideals of equality, tolerance and liberty that came to characterise European political and philosophical thought during the Enlightenment. Modernity and ideas about 'infrahumanity' are closely linked (Gilroy 2000: 54), because modernity was actually a European project, formed as Europeans distinguished themselves from others whom they saw as less, and even less human, than themselves. The concept of modernity – or civilisation – depended on its opposite, the backward, the uncivilised or the savage, categories which generally encompassed those defined as non-European or non-white.

The ideals of modern Enlightenment and liberal thought trumpeted 'liberty, equality, fraternity', to cite the slogan of the 1789 French Revolution, or the idea that 'all men are created equal', in the words of the American Declaration of Independence of 1776. This ideal of a common humanity, supposed to encompass everyone and endow them with equal natural rights, actually depended on divisions between those seen as fit to enjoy such rights and those who were not properly able to take on the rights and the responsibilities that came with them, a category that included women and many non-European or non-white groups, all seen as less able to govern themselves (Mehta 1997). Thus, for example, in the Americas, women, the poor and the illiterate – with the last two categories including most non-white people – were regularly excluded from voting rights well into the twentieth century (Engerman and Sokoloff 2005). In the United States liberal democracy was for a long time seen as perfectly compatible with racist immigration laws, which excluded those deemed racially undesirable (FitzGerald and Cook-Martín 2014).

Scholars of what is known as coloniality – who are mainly Latin Americans – are insistent on this link between race and modernity, which they trace back to the conquest of the Americas. Modernity only became possible through colonialism and the exploitation of non-European peoples; coloniality is the 'dark side' of modernity, the other face of the same coin, the inevitable and necessary counterpart to modernity. Coloniality is more than just the formal structures of colonialism, it is a set of structural and ideological relationships, in which the West, and in particular Europe, dominates and exploits other regions and peoples, with or without the formal structures of colony or empire. (Decoloniality is the project of fundamentally altering these structures.)

Race is integral to coloniality, because it justifies the exploitation and expendability of the exploited peoples. In the Americas, for example, the Spanish decided early on that the natives were human and thus the same as them, but they also inserted 'the colonial difference' by legislating that they were childlike and less rational, and thus should have fewer rights (Mignolo 2011). This, for Mignolo, was a foundational moment in 'the racial classification of the world'. It is a racial difference, even if the language of race is not used: the ranking of humans using a measure based on 'the ideals of Western Christians' is what makes it racial. 'Racialization does not simply say "you are Black or Indian, therefore you are inferior". Rather it says, "you are not like me, therefore you are inferior"'. Racism encompasses physical characteristics, but also religion and languages (Mignolo 2005: 17). This current of thought adopts a broad definition of race, seeing it as constitutive of modernity and, although varying in its form, as a persistent axis of difference and inequality from the fifteenth century to the present day.

Race in Western antiquity

We can see, then, that many scholars link race and racism to a Western-centred history of modernity and colonialism, seeing it as a powerful tool in the armoury of domination and exploitation. Some scholars, however, argue that racial or 'proto-racial' concepts existed well before modernity, whenever that is said to have emerged (Eliav-Feldon, Isaac and Ziegler 2009). As we will see in more detail in the next chapter, historians debate whether Greeks, Romans and others in antiquity (roughly eighth century BC to seventh or eighth centuries AD) thought about human difference and those people seen as 'other' in ways that should be considered racial; particular attention is paid to views about black people. As might be expected, a lot depends here on how historians define race and racism.

Various historians think that ancient texts and images show little evidence that ancient Greeks, Romans and Egyptians – and we are talking here about several thousand years of history, although the main focus is on ancient Greece and Rome – had a negative view of black people (often referred to as Kushites, Nubians or Ethiopians) (Hannaford 1996; Snowden 1983; Thompson 1989). Snowden contends that, during this period, people accepted slavery as normal practice, but while many black people were slaves, most slaves were not black; people were ethnocentric and judged others in terms of their own group values, for example in relation to standards of beauty; and they often assumed people not like them were barbarians. But none of this amounted to racism: there was no hint of 'biological racism', black skin was not taken as a sign of inferiority and skin colour was not an obstacle for integration in either Greek or Roman society; colour was not used as a basis for judging people (Snowden 1983: 63–79).

Thompson (1989) finds more evidence of an ancient Roman aversion to people of black African appearance, preferring people who looked like themselves, and he thinks that Roman attitudes did tend to associate blackness with low status (not surprisingly, as most black people in ancient Rome were of low status). But all this still falls short of racism, in Thompson's view, as the relationship between physical appearance and status was not systematic and predictable, nor did it involve the systematic exploitation of one group by another: 'Roman attitudes to *Aethiopes*, even at their most negative, have nothing to do with the familiar modern phenomenon of race' (1989: 157). It was more a question of ethnocentric aesthetic tastes and reactions to unfamiliar people. Focusing on early religious texts, Goldenberg likewise finds no negative images of black people in the Bible itself, and also concludes that in the ancient Jewish world up until about the seventh century AD colour was 'irrelevant in taking the measure of man' (2009: 200).

The definitions of race and racism are, of course, crucial here. Goldenberg, Snowden and Thompson (and Hannaford 1996: 58) operate with what might be called quite a modern concept of race, involving systems of domination, based on ranked hierarchies of groups defined as unalterably different in biological and cultural terms. Isaac takes a different line, and finds evidence for at least 'proto-racism' and even racism in classical antiquity. For him, racism involves an attitude towards individuals and groups of peoples, 'which posits a direct and linear connection between physical and mental qualities'. 'It therefore attributes to those individuals and groups of peoples collective traits, physical, mental, and moral, which are constant and unalterable by human will because they are caused by hereditary factors or external influences, such as climate or geography' (Isaac 2004: 23). On this basis, racism regards individuals and groups as inferior or superior. Isaac concedes that the Greeks of the late fifth and early fourth centuries BC were not even proto-racist, but thereafter and also in Roman society ideas developed that he sees as proto-racist. According to him, human difference was explained largely in terms of the influence of the environment: groups of people were the way they were because of characteristics engrained into them by the climate and the physical environment in which they lived (which might include astrological influences). These traits were absorbed into the body and passed on through heredity. Overall, environment and heredity combined to create an outlook 'almost as deterministic as modern racist theory' (2004: 503). Thus people who were conquered and turned into slaves would, over time, pass on their slave status and eventually become what Aristotle called 'natural slaves'. Isaac also notes that ancestry was seen as important in shaping a person's qualities and essential attributes and that the ancient Greeks were concerned with keeping lineages strong and pure, avoiding degeneration through good marriages and some use of eugenic practices (see Chapter 3).

The historical evidence is voluminous and contested, and I cannot summarise it here. The important points are (a) that in the ancient Western world there are certainly traces of styles of thinking about difference that resonate with what we recognise as 'racial' thinking: whether one calls them racial or not depends a great deal on the exact definition of the term, as well as the interpretation of detailed evidence; and thus (b) that the delimitation of 'race' as a domain of enquiry is artificial, that is, an analytic artefact or something we construct for analytic purposes, rather than something that is simply 'out there' as an objective fact. It may be restrictive to link race to modernity, given what we know about the ancient world, but I think it still makes sense to do so in order to delimit a specific domain of enquiry.

Is anti-Semitism a form of racism?

Another problem for the view that links race to modernity is the question of the Jews. If race emerges with colonialism and Western domination of the rest of the world, does this mean anti-Semitism is not a form of racism? This seems counterintuitive, to say the least, when in the nineteenth and twentieth centuries anti-Jewish prejudice has been one of the most virulent forms of racism, particularly *within* modern Europe (Frederickson 2002). The Holocaust stands as a horrifying testament to the power of racism.

The answer to this problem actually overlaps with the issue of racism in pre-modern times, because anti-Jewish prejudice goes back into early medieval times at least, and again the question emerges of whether these early forms of prejudice can be classed as racial. This raises a related question: if being Jewish is at one level a religious status, in what sense can it be seen as racial (Nirenberg 2009)?

In fact, the overlap between religious and racial intolerance, and anti-Semitism specifically, are vital to grasping the emergence of racial thinking, as we will see in Chapter 2. In Christian Europe anti-Jewish sentiment can be traced back to the fourth century AD, and increased markedly over the centuries: Jews faced pogroms and in the end expulsion from Spain in the fifteenth century (Thomas 2010). Conversion to Christianity was always an option for Jews, but especially by the fourteenth century the fidelity of new converts was often seen as suspect, and Jews (and Moors) were discriminated against on the basis of their Jewish and Moorish 'blood' or ancestry (which the Spanish by then called *raza* or race), seen as impure. This indicates a form of thinking about difference that hovered between religious and racial criteria. Blood, filiation (the connection between parents and children) and ancestry were a key idiom for thinking about religious differences; and religion, or rather religious intolerance and conflict, was a formative terrain for the emergence of ideas about race, understood as ancestry, descent and blood.

The conclusion to be drawn is that the ideas of race associated with Western modernity and its colonial projects drew heavily on the medieval concepts of blood and purity that circulated in Iberia and that were intimately linked to religious differences. These ideas formed one basis for emerging forms of racial thinking in the Americas, where African and indigenous American blood were also seen as tainted and dishonourable, even if African and indigenous people were Christian converts (Wade 2009b: 67–9). The New

World colonies were a vital context in which Western racial thinking developed, encompassing not only differences and relations between Europeans and their colonised or enslaved subjects, but differences among Europeans too. Much later Jews became defined as an innately inferior biological race – above all in late nineteenth-century Germany – well after such biological racism had been developed in European science and in the US context of black–white relations (Frederickson 2002: 72). Jewish intellectuals themselves deployed the idea of Jews as a distinctive race in the late nineteenth century, with central importance given to the idea that Jewishness was transmitted in the blood, through the maternal line (Abu El-Haj 2012; Hart 2011). The idea of a Jewish race persisted, appearing in 1930s discussions about Jewish and Arab conflicts in Palestine and, most notoriously, in Nazi racial theory. Some argue that the idea of a Jewish race lies behind current Israeli discrimination against Palestinians and other Arabs. The fact that some 40 per cent of Israelis are Arab Jews complicates the idea of Israel as a European outpost and 'presumptively white' nation, and lays more emphasis on the idea of Jewish ancestry as the defining criterion for inclusion (Goldberg 2008: 116, 140).

Thus Jewishness (or rather, religious conflicts involving Christians, Muslims and Jews) was a vital ingredient in the early development of racial thinking, and later became powerfully racialised by the very modes of thought that developed from seeds planted in this early context. Overall, then, it continues to make sense to focus on race as an integral aspect of Western modernity. What the case of the 'Jewish race' indicates is that, once developed, racial thinking can colonise and redefine other kinds of difference.

CONCLUSION: SO WHAT IS RACE?

What can we conclude about race, after this whistle-stop anthropological and historical tour through the chronology of race, comparisons with India, China and the Amazon, and historical excursions into pre-modernity and Jewishness? The purpose of this review has been to show that race has two interconnected dimensions: one linked to the kind of classificatory criteria that characterise racial thinking; the other to the historical presence of the phenomena of race and racism.

1. The first of these dimensions – the classificatory criteria – involves a focus on the ways people differentiate people into classes or categories, using ideas about bodies, blood, heredity and filiation. This is similar to, but a good deal broader and more complex than, the approaches that say race is about the categorisation of 'physical appearance', 'phenotype' or 'biological difference'. The following are key points:

   - Classes and categories are rarely sharply defined; instead, racial categorisation, like most human categorisation, depends on typical examples or stereotypes.
   - Bodily differences of physical appearance, when they become significant, tend to focus on differences seen as spread across a culturally meaningful geography; that is, some types of physical difference are made racially meaningful and not others.
   - Bodily differences may be understood as mainly internal, 'in the blood' or inside the body, and not necessarily visible.

- Bodily differences are seen as linked to heredity, creating perceived lines or lineages of similarity across the generations.
- This makes sexual reproduction and thus gender relations intrinsic to ideas about race (Collins 2000; Nagel 2003; Wade 2009b).
- Differences in bodies are always perceived as linked to behaviour, thought and moral qualities, without the former necessarily being seen to fully determine the latter; race is always about both 'nature' and 'culture'; it is a 'natural–cultural assemblage'. In some contexts, differences that Western philosophy or social science might establish between nature (an underlying biology that is fixed) and culture (an overlying learned and malleable accumulation of habits, behaviours and ideas) do not operate, such that 'cultural' features are seen as 'natural'; this means that we cannot say that 'race' is defined in terms of its reference to 'nature' (if this is defined as an underlying biology), because in some contexts what we would call culture is considered natural. In eighteenth-century thought a man's moral qualities might be seen as being as natural as his physique. In late twentieth-century thought 'culture' might be seen as 'almost biologised' (Gilroy 1987: 61). This means we need to understand local concepts of nature and culture, or more generally, how people are constituted as embodied persons.
- The previous point means that aspects of appearance that are not strictly phenotypical can become racial markers: beards, hairstyles, body adornments and alterations (tattoos, piercings), and even clothes can become racialised.
- Bodily differences, while they may be rooted in ideas about nature or even biology, are not necessarily fixed. The frequent insistence that race is *always* about traits seen as unalterable, immutable, fixed and unchangeable is misleading. Racial thinking may indeed act to fix certain aspects of the person or the group as enduring traits that are not subject to change, but it may also be about *changing* bodies and *changing* behaviours. Thus, in a variety of contexts, from early colonial Latin America to late nineteenth-century Dutch colonies in South East Asia, colonists feared that their physical constitutions, as Spaniards or white Europeans, could weaken and be lost in the local environment.

2. All the above features can be found, in some form, in situations that we do not necessarily want to encompass in the analytical field of 'race'. As we have seen, many of the criteria listed above can be found in pre-modern Western or non-Western contexts. The criteria outline a toolkit of modes of thinking about human difference, based on ideas about bodies, filiation, heredity and blood; these modes are widespread, and racial thinking uses them in flexible and selective ways, but if we include all occurrences of these modes as racial, the term quickly becomes analytically very baggy. Therefore we need to add a second dimension, which is to understand race as closely linked to Western projects of modernity and domination, understood as beginning with European expansion into the Americas.

   - This brings a concern with inequality and domination into our understanding of race, not as a necessary criterion (which would say that, if we see the criteria listed in (1) above, plus inequality, we are necessarily looking at 'race', irrespective of the context), but as a historical fact, linked to a specific history of Western projects of modernity.

- This is an analytical artefact, a decision made to limit the field in a specific way. We have seen that other contexts have features that are associated with race; they are race-like or proto-racial. But, while we can see some of these as having fed into the historical phenomenon that we label as race – such as ideas about Jews and Moors in medieval Spain – I argue that we do not want to include them in the field of race.
- The motivation for this is that race should capture a specific historical trajectory, in which the word race itself emerged and the concept developed in ways that shaped world history.
- This does not close off the definition of race, because, as it has changed meanings in the past, it continues to change in the present, and will doubtless change in the future. Racism is a 'scavenger ideology' (Mosse 1985: 234), which picks up ideas from surrounding contexts and gives them racial meanings and uses them to enact racist exclusions.
- The changing form of race also means that we need to avoid the implication, found in the coloniality approach, described above, that race and racism suddenly came into being in the fifteenth century, as a justification for Western colonial domination, and continued more or less unchanged thereafter. It is important to trace the variations that race undergoes as it adapts to changing contexts.

These two broad dimensions need not always be explicitly present together. One may evoke the other, as when talk of 'blacks' and 'whites' (colonially derived categories) connotes ideas of bodies, descent and heredity. In this sense, race is 'relational': it is made up of related elements, some of which may be hidden but which are suggested by the presence of others (M'charek 2013).

On the basis of this definition, especially the criteria listed in (1), in Part I of this book I will challenge the standard chronology of race with which I started this chapter. I do not reject it out of hand, but I do think that it is misleading to talk in terms of a broad trajectory which leads from a concept of race built around 'culture' (fourteenth to eighteenth centuries), to one built around 'biology', seen as immutable (nineteenth century and first half of the twentieth), to one built again around 'culture' (after WWII). This seems to me to obscure the fact that race is always about both 'nature' (or 'biology') and 'culture' – and that we need to be clear about what both of those concepts mean at different times.

FURTHER RESEARCH

1. Try the exercise at the beginning of this chapter with different people (older, younger, different nationalities, etc.) to see what connotations the idea of race has for them.

2. Enter 'race and ethnicity' as a search term in Google images. What kind of ideas about race emerge from the images that appear and how do these ideas relate to the themes in this chapter?

PART I

RACE IN TIME

2 Early approaches to understanding human variation

In the previous chapter we were concerned with outlining what 'race' means in a way that grasps its variety, while also placing some constraints on that variety to avoid placing too many very diverse eggs in one basket. Some of the features that are often said to define race – the use, by people classifying people, of criteria of physical appearance, nature, heredity, bodily form, biology – can be found in many contexts, lending them some affinity with contexts more typically associated with the idea of race. But it seems unhelpful to include all these very heterogeneous contexts together. Instead, the history of Western modernity, broadly defined, can be used as a thread to string together a specific set of contexts in which race is most helpfully understood as a specific phenomenon.

That thread, however, has to be seen as itself diverse. The usefulness of 'Western modernity' as a connector or umbrella should not tempt us into assuming that ways of thinking about physical appearance, nature, heredity and bodies do not undergo very significant changes over time; not to mention the ways of perceiving habit, customs, behaviour, civilisation, morals and culture that always go alongside them in racial thinking. Nor should we assume that the thread has a simple starting point: as I briefly indicated in the previous chapter, the ancient and medieval worlds had plenty of conceptual elements related to all these ways of thinking, even if we decide not to label these as racial.

This chapter explores how ideas about what we can, for shorthand purposes, call 'nature' and 'culture' have worked together in ways of explaining humans and human diversity. The aim is to describe the conceptual resources – about bodies, environments, heredity and behaviour – that were available early on for an emerging mode of racial thinking, and how these developed in the context of European colonialism. The argument is that ideas about bodies, environments, heredity and behaviour were constantly intertwined in historically changing ways in this overall context and that it makes sense to grasp all these ways as 'racial', rather than confining the idea of race to, say, the Enlightenment and thereafter. By the same token, the argument is that, while modes of racial thinking certainly changed over time, it does not make sense to think of pre-eighteenth-century modes of racial thought as basically about 'culture', with the corollary that after this time racial thought was mainly about 'biology'. When Todorov (1993: 101) says of late eighteenth-century French ideas about human diversity that 'there is no reason as yet to speak of racialism: we are operating within a classification of cultures not bodies', I think he is making a double error: first, it is wrong to oppose 'racialism' to 'cultures', because race always invokes both 'nature' and 'culture'; second, I think there is plenty of evidence that European thinkers at this time, and earlier, were very concerned with bodies and their classification, even if they employed different classificatory concepts and methods from those used in the nineteenth century.

2.1 Nature and culture

The emphasis in this chapter is on history, but the approach is identifiably anthropological. The influential anthropologist Claude Lévi-Strauss defined 'the central problem of anthropology' as being 'the passage from nature to culture' (1963: 99); he was posing the question of how people make culture(s) out of the raw material of nature, including their own human nature. In pursuing this question, anthropologists have realised that 'nature', and especially human nature, are not simple baselines that anthropologists can understand as a resource or raw material. Along with historians and philosophers, anthropologists have been pioneers in exploring the variety of people's ideas about the constitution of humans and their surroundings. They have shown that different peoples have different viewpoints on what 'nature' is; they do not necessarily subscribe to a Western version of the concept in the first place. As we will see, the idea that there is a single 'Western' concept of nature (or culture) is not actually sustainable, as ideas have varied over time, but for the purposes of our discussion it is useful to outline some core ideas about nature and culture that have been current in the West since about the nineteenth century, with important roots in earlier periods (Soper 1995; Strathern 1992; Williams 1988).

These ideas maintain, first, that nature and culture are two distinct and separate domains. Nature, including human nature, is a given, underlying set of resources, which are also constraints; it is pre-social, instinctive, mechanical (in the sense that it is driven by complex mechanisms or the 'laws of nature') and biological. Culture is a separate domain which works on, builds upon and colonises nature; it is a human activity and, although humans are 'naturally' predisposed to acquire culture, they have to do so through a process of 'nurture', learning and socialisation. Culture emerges as a superstructure built by humans on top of, and constrained by, the basic raw material and resources of nature, including their own nature. This nature–culture opposition resonates powerfully with other important oppositions in Western thought, such as that between body and mind (often associated with the philosopher René Descartes), or between individual and society (often linked with the sociologist Émile Durkheim, for whom the individual started life as a natural entity and was nurtured into a full member of society).

In other contexts, ideas might be rather different. In Papua New Guinea, for example, people recognise a distinction between the wild and the domestic. This resonates with some aspects of a Western opposition between nature and culture in referring to things that occur without human intervention and things, including social relationships, made by humans. But it is very different in that the wild is not 'nature', construed as a set of given resources and constraints in the human body and in the environment, and the domestic is not 'culture', understood as human processes of making things and social institutions that build upon and transcend nature. For New Guineans the wild is a domain of non-human powers and things, but it is not a set of innate givens in the environment, and humans are not seen as having an underlying given nature, which has to be socialised or trained. Humans are not divided between a natural component and a cultural one: everything human is domestic or social, but the domestic realm is not seen as one that emerges from a taming or domestication of the wild (Strathern 1980).

Even in the West, anthropologists have pointed out that the apparently simple distinction between nature and culture is actually not straightforward. The distinction is vital in part because it is foundational for science, which depends on the idea that there is an objective realm of nature, of given facts that can be revealed through controlled observation and experiment. Culture is the realm of beliefs, values and ideas, which may be factually right or wrong, or are simply not facts at all (e.g. they are moral judgements about what is good). In science, culture has to be held apart from nature, so that the discovery of objective facts is not distorted by beliefs and values. But in everyday life people do not police the boundary so vigorously, and it often becomes unclear whether something is counted as natural or cultural. As we saw in Chapter 1, in cultural racism culture can become 'biologised' and presented as natural. Edwards' ethnography of a northern English town also shows that people shift ambiguously between things that are said to be 'born' in people and those that are 'bred' into them by upbringing (Edwards 2000). (Indeed, in English the very notion of 'breeding' mixes ideas of nurture and nature: 'breeding pigs' implies controlling their upbringing but also their reproduction.) Even scientists may not separate nature and culture as rigorously as they might think: the history of race in science is itself a testament to this, as the 'scientific facts' about race were strongly conditioned by beliefs and values about race, as we will see in Chapters 3 and 4. More broadly, the anthropologist of science Bruno Latour argues that scientists routinely mix nature and culture, as this is the only way they can make scientific projects actually work in practice: 'finding the facts' requires a complex assemblage of people, money, institutions, technologies, laws and social relationships (Latour 1993).

An anthropological sensitivity to the fact that recent Western ideas about nature and culture should not be taken for granted is important for an understanding of race. If this concept involves a process of 'naturalisation' – making reference to bodies, heredity, biology or human nature – then we need to know what that process implies in terms of what 'nature' means in a given context. And if race always involves nature and culture working together, then it is important to understand what these concepts mean and how they are thought to interrelate in different contexts.

2.2 Ancient Greece and Rome

In the previous chapter I briefly outlined debates about whether 'racism' could be said to exist in ancient Greece and Rome: a lot of attention has been paid by scholars to Greek and Roman attitudes to black Africans. In this section I will look in more detail at evidence about this, but the main aim is to put the material in the context of contemporary ideas about nature, to see what naturalisation meant at that time and what kind of conceptual elements the Greeks and Romans provided for later ideas about human diversity and judgements about the qualities of people perceived as different.

Conflicting interpretations

The consensus among recent historians is that ancient Greeks were not racist in the 'modern' sense of the term, when this is understood as involving 'virulent colour prejudice', 'biological racism', or skin colour being 'an obstacle to integration in society' (Snowden 1983: 63). Snowden argues against interpretations that see evidence of anti-black sentiment in the ancient world. He says that black Africans were a familiar sight in the ancient Mediterranean world, frequently as soldiers (whether as warriors defending their homelands or as mercenaries fighting in other armies) and as prisoners of war and slaves. From the ancient Egyptians through to the Romans, the image of the Nubians, Kushites and Ethiopians was 'remarkably similar and consistent', being 'highly positive' (Snowden 1983: 37, 59). Many black Africans in Greece and Rome were slaves, but most slaves were not black. Nubia as a region was not associated with backwardness or cultural inferiority. Black skin colour was noted, but not attributed much importance. Greeks and Romans generally gave higher aesthetic value to a body image that resembled their own (or an ideal version of it), but dark skin might also be praised, for example in poetry. Black Africans were represented in visual images (mosaics, pottery, etc.) in varied ways: there were some satirical portraits (as there were of whites), but also sympathetic ones. Although there was a tendency to associate the colour black with evil and ill fortune – and dark-skinned people might therefore sometimes be linked to bad omens, at least by the Romans – this was a minor trend.

Black Africans living in Greece or Rome were often enslaved prisoners of war, but there are also examples of black people who achieved some upward mobility, if they assimilated culturally; they encountered no exclusions by virtue of their colour alone. There was also no social pressure against marital or sexual unions between blacks and whites. Human diversity, including physical appearance, was explained by environmental theories, which were 'unprejudiced', because they were based on the idea that all humans were fundamentally the same (1983: 87).

Hannaford agrees with Snowden, but focuses on a different aspect. For him it is crucial that, among the Greeks, 'we do not find Nature independent of ethics and morals' (1996: 58). The Greeks did not try to divide humans up into types, according to their material, natural characteristics. Much more important was the form of political governance and civil association. The Greeks differentiated between themselves and 'barbarians', but this distinction was based not on geography, appearance or ancestry, but on how people lived. Barbarians were those who lived according to the laws of *physis* ('nature' in the sense of the domain of things that exist independent of human intervention) and thus lived brutishly, inarticulately, without real choice, governed by custom and habit alone, bound together by descent, and subject to tyranny and hierarchy. All people had to contend with *physis* and custom, but they did not have to be ruled by them alone. Those who lived by the rule of *nomos* – man-made laws operating through reason, moderation and properly formed speech in a public assembly – were civilised citizens who lived in the political way. People who lived under the rule of monarchs and despots would be inferior. Slaves, who did not have autonomy and free time, could not aspire to live under *nomos*. Some people were 'natural slaves', because although they were born rational, like all humans, they did not have the full range of rationality needed to be a proper citizen.

Other writers put a good deal more emphasis on Greek theories about the environment and the way it shaped people, collectively and individually (Glacken 1967; Goldenberg 1999; Isaac 2004). This did not add up to a racial theory, in the sense of dividing humans up into permanent physical types, with associated moral qualities, but it does add another dimension to Hannaford's emphasis on the political way of life.

Isaac takes a strong line here, and argues that the Greeks from the late fifth century BC espoused a form of environmental determinism in which the 'collective characteristics of groups of people are permanently determined by climate and geography' (2004: 163); some writers attributed great importance to the political climate, others less so. Aristotle argued that northern European climates produced people who were free spirited but not very bright, while Asian climates produced people who were soft, intelligent and servile. The Greeks constituted a happy medium term, which suited them to rule over others (2004: 71–2). Aristotle's defence of 'natural slavery' – which he made in an inconsistent fashion – implied that some people were born inferior, a concept certainly disputed by other Greek thinkers. Because of their tendency to be servile, Asians were more likely to be natural slaves, whereas Greeks were not natural slaves (2004: 178). In addition, by the fourth century many Greek thinkers regarded the Greeks as a people 'uncontaminated by an admixture of foreign elements, and ... therefore superior' (2004: 165).

The permanence Isaac attributes to collective characteristics may be overstated. As he recognises, all these thinkers believed in the inheritance of acquired characteristics – that is, traits that a person acquires in their lifetime are passed on to offspring. For Isaac, such a belief 'paradoxically' leads to the idea that traits become permanent in a group, because they are hereditary (2004: 164). But the belief also implies the possibility of change: if people change environments, they can adapt and pass on those changes to their children. Isaac too readily assumes, from generalising comments about Asians and Europeans, that all their characteristics were necessarily permanent. While recognising that the Romans, if not the Greeks, integrated many non-Romans into their society, he does not have much to say about this process of integration and how it squares with the supposed permanence of collective characteristics.

If we look at Greek theories about heredity it appears that, while there was an expectation that the mechanisms of sexual reproduction would produce offspring that were recognisably like their parents, especially their fathers, there were also other factors that introduced change at this individual, inter-generational level. Climate and diet could affect the father's semen and the mother's menstrual blood – which were held to shape the embryo – and thus the appearance and moral qualities of the offspring. The mother's experiences – things she saw, things she ate – during pregnancy could also affect the physical and mental constitution of the child (Stubbe 1972). These are 'soft' theories of heredity, which envisage the possibility that the material passed from one generation to another might change in the process (unlike genes, which although they are reshuffled during sexual reproduction do not themselves change). Environmental theories, then, are always by definition theories about change as well as determination.

Lessons from the ancient world

What can we learn from this material about the ancient world, in which we see such diametrically opposed interpretations as those of Hannaford and Isaac? It is worth noting that the way they read the evidence depends in part on how Greek ideas are seen to relate to later developments. Hannaford insists that later thinkers from the Renaissance onwards selected bits from Greek texts that fitted with their ideas about race, thus casting thinkers such as Aristotle and Plato as proto-racist and misunderstanding the real character of Greek ideas (1996: 59). Isaac thinks that Greek and Roman theories about how the environment could shape the constitution and character of whole peoples provided the basis for later, more clearly racial and racist ways of thinking about human diversity.

For our purposes the key points are, first, that the debate about whether the Greeks or Romans were not racist or were proto-racist is both important and arbitrary. It is important because it helps historicise race: racial ways of thinking have a distinctive history, and what we understand as race and racism have not always existed. It is arbitrary because the exact point at which we can say that race or racism actually came into being is impossible to pin down. This is why I emphasised in Chapter 1 that confining the study of race to the period of Western modernity – the start of which is itself impossible to pin down – is an *artificial* decision: it is useful analytically.

Second, among the Greeks we find ideas about human diversity that link people's physical and moral constitution to environment, especially in a collective way. Although not necessarily racial in themselves, such ideas are a key component in the toolkit of racial thinking: they form a naturalising argument, which has proved long lasting and adaptable. Groups of people have certain constitutional characteristics – of body, temperament, capability, etc. – which are engrained in them by virtue of their external surroundings, which they share, which form part of their heredity, and which make them distinctive. Surroundings need not be only climate and geography; they can include social context. The point is that the surroundings are thought to shape the very constitution of the person: for the Romans, being enslaved eventually made people slavish by nature. (Evaluating these constitutional characteristics as good or bad is, of course, another step, but one which is frequently taken.)

Third, what we would now call natural (or biological) and moral–cultural traits are mixed together: a person's 'nature', in Greek terms, could include all kinds of traits we would see as cultural. Some interpretations of racial thinking say that the assumed link between physical and non-physical characteristics is 'a crucial component – in fact, the lifeblood – of racist thinking' (Goldenberg 1999: 562), but this assumes that we already know what is understood to be 'physical' in a given context. Racial thinking frequently has used specific aspects of physical appearance as a marker, but it is important to see that such thinking is not confined to the use of these markers: judgements and assumptions about a person's or a people's 'natural' constitution can be made without resorting to physical appearance, as the Greek material shows.

Fourth, while this way of thinking about human diversity naturalises and envisages human constitutions as engrained and hereditary, there is also always some flexibility and room for change. This is the reason I use the word engrained, rather than fixed, permanent or immutable. Engrain originally meant to dye, a permanent process – until the dye fades, is bleached or is superseded by another dye. Environmental explanations, ideas about the inheritance of acquired characteristics and 'soft' theories of heredity all leave space for characteristics seen as engrained to be changed: the constitutions of persons and peoples can alter within a generation or over several generations. That is, racial thinking does not necessarily rely on ideas about immutability and permanence; it includes, for example, ideas about degeneration and improvement, frequently as a result of changing surroundings.

None of this adds up to an argument that the Greeks were, or were not, actually racist. Such a judgement depends on one's definition of race and racism. Hannaford (1996: 58) uses a restrictive definition of race, based on the following ideas: that people are like animals and can be explained in terms of the laws of nature; that descent is about the transmission of biological characteristics; that people are classified into racial types, arranged hierarchically, based on observable structural characteristics and regarded as real entities. Isaac uses a looser definition, emphasising judgements of inferiority and superiority based on ideas about groups of people sharing traits thought to be caused by hereditary factors or external influences. I think Isaac is right to show that the Greeks and Romans made some judgements along these lines – which for him constitutes 'proto-racism'. The important point is that these modes of thought fed into later ways of thinking about human diversity; in my view, these ways become 'racial' when they articulate with European colonialism and modernity, defined in a broad sense.

2.3 Medieval and early modern Europe

As indicated in the discussion about anti-Semitism in Chapter 1, late medieval Europe and particularly Spain were a crucial context for the development of thinking about 'blood' and religious belief and practice, which was then transmitted to the New World and the encounters of Europeans with Africans and native Americans.

The importance of blood, ancestry and genealogy needs to be placed in the context of European family and inheritance arrangements. In broad terms, European economies were based on plough agriculture, which men controlled. Society was stratified, with important families holding landed estates as a key basis for their wealth. Property was passed on to both sons, directly, and daughters, through dowry. As significant amounts of property passed via daughters, particular importance was attached to their marital alliances, which were vital to controlling property. The legitimacy of offspring was also crucial to making sure family property was kept within certain bounds: the virginity of brides and the chastity of women in general – especially among the landed classes – therefore took on great importance; female sexuality and 'honour' were strictly controlled by fathers, brothers and husbands in a patriarchal social order. These concerns about sexuality were part of a preoccupation with genealogy and with tracing the lines of descent and alliances that defined nobility and regulated status and wealth; this included ideas about the purity of genealogical line, defined in terms of a continuous succession of ancestors considered appropriately noble. Such a preoccupation was evident among the elite; plebeian people had less property and thus less concern with genealogy (Gies and Gies 2010; Goody 1976, 1983).

Genealogy and ancestry are closely linked with the emergence of the term race itself in European languages. Predictably for a concept that has varied meanings, there is uncertainty about its etymological origins. Some argue that it derives from the Latin *ratio*, literally meaning calculation or account, but also manner or system, which by the Middle Ages had acquired the meaning of type, kind or species. Others argue that it derives from words in both Arabic (*ra's*, meaning head) and French (*haras*, stud, i.e. animal used in breeding) and thus refers to lineage or pedigree, originally in relation to horse and cattle breeding. The term *razza* appeared in Italian as early as 1300 (with reference to horses). It also appeared as *rassa* in the southern European *langue d'oc* in poetry of the early 1200s, referring to the character of a set of people. By the 1430s it appeared in Spanish (*raza*) as a synonym of lineage or ancestry, and then appears in French and English in the sixteenth century (Corominas 1980; Liberman 2009; Nirenberg 2009: 248–50; Voegelin and Vondung 1998 [1933]: 80–3). Although the etymology is unclear, race is connected both to ideas of lineage and to types of living things (human or otherwise): indeed, in Spanish, French and Italian the word is still used to refer to breeds of domesticated animals and plants (e.g. dogs, horses, cows, maize). The etymology suggests that race referred to categories of living things that share a character by virtue of common ancestry.

Genealogical reckoning is central to emerging ideas about race, and brings with it an important gender dimension. As noted above, where legitimacy of offspring and 'purity' of lineage were concerned, it was women's sexual behaviour that was a focus of preoccupation; men's 'honour' was not considered tainted by sexual promiscuity in the same way. Women's subordinate status in medieval Europe had many dimensions, but a vital one was the control of their sexuality by men. As we will see, this implies that a concern with racial ancestry would very often be expressed in terms of a concern with sex and gender relations.

Jews, Moors and 'clean blood'

The concern with ancestry and the use of the word and concept of race took a specific turn in the context of religious conflict in Spain in the fifteenth and sixteenth centuries. As outlined briefly in Chapter 1, this was connected to Christian ideas about Jews (and also Moors or Muslims, originally those from North Africa), so it is useful to give a little background to these ideas, partly because, although the focus below is on Spain, the basic concern with blood and status was more widespread, as was a preoccupation with Jewishness.

Jews were a problem for Christianity because they were blamed for killing Christ and they did not accept him as the Messiah. They were seen as a collective unit, marked by their faith as different. Laws banning Jewish–Christian marriage were passed by the Roman emperor in 388. But Jews could convert – indeed, Christian theology held that they should be converted – and thus be assimilated. By the tenth century, in an increasingly

Christianised Europe, Jews were subject to ever more stringent control, being gradually restricted to the occupation of money-lending (usury), seen as sinful. By the thirteenth and fourteenth centuries attitudes became more hostile, involving ideas about Jews as linked to the Devil, as tainted by infected blood, and as having defective or unclean bodies. Stories circulated about Jews sacrificing Christian children or poisoning wells during the plagues of the Black Death (1348–50). The infidelity of Jews was now seen as organic, in the body and the blood, and intrinsic. Such beliefs were accompanied by intermittent massacres and pogroms (Frederickson 2002: 18–25; Thomas 2010).

As discrimination increased, in the context of growing conflict between Christians and Muslims in Spain, Jews faced pogroms in 1391 and finally expulsion from Spain in 1492, when the Moors were also finally driven from the Iberian Peninsula. Many converted and became 'New Christians', and, although some Christians defended the power of the holy spirit to transform the convert, body and mind, into a true Christian, many suspected the real fidelity of these newcomers. From 1449 laws were passed banning people who did not have *limpieza de sangre* (cleanliness of blood) from certain occupations (e.g. public office, the priesthood, many guilds and universities). Clean blood meant being without *raza de judío o moro* (race – in the sense of ancestry – of Jew or Moor). Jews were said to have *sangre infecta* (infected blood) and as early as 1438 the writer Martínez de Toledo had argued that 'the wretched man, of vile race or lineage, no matter how great or how wealthy, will always return to the vileness from which he descends'. Later, the writer Juan de Pineda said in his *Agricultura cristiana* (1589), a catechistic manual, 'no sane person wants a woman with *raza* of a Jew or a *marrana* [lit. pig, a term applied to a newly converted Christian, who continued to practice Judaism in secret]' (Corominas 1980, entry on 'raza'). In that sense, conversion did not fully work. Blood or race would stubbornly express itself.

Yet these blood purity laws were not always rigorous: to prove freedom from the taint of Jewish or Moorish blood, a person commonly had only to provide details going back two generations (Martínez 2008: 47; Twinam 1999: 43); in 1550 Charles V said that the children of Jews and Muslims should not be excluded from church office, saying: 'Sons and descendants are to be admitted if they are good Christians and capable and competent' (Poole 1999: 364). In the spirit of Christian universalism, some people defended the new converts. Over the sixteenth century, however, the importance attached to *limpieza* increased and the limitations on the depth of genealogical proof required were eroded, meaning that even the slightest 'taint' of Jewish or Moorish blood could disqualify a person. For some purposes, a man had to show that his wife was also pure.

By 1600 the whole concept of *limpieza* had become more 'essentialist', separating pure or Old Christians from New Christians, rendered unclean by Jewish or Moorish blood, in a 'more rigid dual-descent model of classification'. At the same time, greater scrutiny was directed at women. New Christian women were suspected of secretly practising Judaism in the domestic space and instilling it in their children; and the breast milk of women of Jewish or Moorish descent who acted as wet nurses was also feared as a potential contaminant for Old Christian children. Meanwhile, the sexual behaviour of Old Christian women was controlled more tightly, for fear they could introduce impurity into the family and the lineage (Martínez 2008: 52–7).

But in all this there was uncertainty about how firmly Jewishness remained in the blood, and how this shaped what people were: 'medieval people had a great many ways of thinking about the transmission of cultural characteristics across generations, such as pedagogy and nurture, which did not necessarily invoke nature, inheritance, or sexual reproduction' (Nirenberg 2009: 255). Heretical beliefs could be passed on through imitation as well as in the blood; or rather, as we have already seen for other contexts, people did not 'articulate a clear distinction between "nature" and "nurture". Rather they tended to attribute the transmission of beliefs and behaviour to both cultural and biological inheritance and to conflate the two'; thus '"natural" traits were by no means rendered immutable' (Martínez 2008: 47, 48).

We can see here a negotiation between religious and genealogical modes of reckoning – could conversion transform Jew into Christian, or would Jewish ancestry stubbornly resist its effects? We can also see a fusion of what we would call biological and cultural attributes: a religious ritual could shape the whole constitution of the person, transforming Jewish race; but Jewish blood might still persist, evident in the secret practice of Jewish rites or transmitted in the breast milk of a New Christian wet nurse.

As with the debate about racism in the ancient world, it is a matter of judgement whether one concludes that medieval Spain was racist in the 'proper' sense of the term (i.e. the sense that became dominant after 1700 or 1800, depending on one's perspective); but it is clear that some key elements of racial thinking – linking bodies, heredity, character – were being developed in a context of religious conflict and intolerance, and being deployed to exclusionary effect. We do not need to restrict the definition of racial thinking to ideas about completely fixed and immutable 'biological' characteristics: racial thinking also works in this more flexible space of uncertainty and change in relation to bodies, heredity transmissions and engrained characters.

Views of Africans and others

If we can find in Spain a particular version of ideas about blood, ancestry and race that will prove important components of developing concepts of race, we also encounter changes in the way people in Europe and the Mediterranean world were thinking about human diversity, building on existing geographic and climatic theories. Influential medieval Islamic writers from the Middle East and Mediterranean had clear ideas about the inferiority of certain peoples. Avicenna, the eleventh-century Persian philosopher, thought that some people 'like the Turks and negroes and in general people living in an unfavourable climate' were 'slaves by nature'. Maimonides, a Spanish-Jewish philosopher (d. 1204), said that people such as the Turks and the Ethiopians were 'irrational beings, not human beings, below humanity but above the ape'. Ibn Khaldun, a fourteenth-century North African historian, stated: 'The only people who accept slavery are the Negroes [Sudan], owing to their low degree of humanity and their proximity to the animal stage.' Such views had precursors among ninth- and tenth-century writers of the Islamic world, who had equated black Africans with animals (Goldenberg 1999: 565–6).

Lewis (1992: 37–42) argues that the context for such beliefs was the formation, from the seventh century, of a vast Islamic empire, which conquered regions in Europe, Africa, the

Middle East and Asia. As Islam expanded, non-Arab Muslims were regarded as inferior by Arab conquerors, who also prized Arab ancestry and, for a time, managed to keep people of mixed Arab and non-Arab ancestry out of the highest political positions. Conquest involved the enslavement of non-Arabs, including the large-scale importation of African slaves, such that slavery became the main way in which Arabs and Muslims encountered Africans. Although Islam as a religion – like Christianity – has a universalist philosophy, which includes all believers with no attention to questions of origin or appearance, Lewis suggests that the reputation of the Islamic world as a place free from racial prejudice is a nineteenth-century myth and that the teachings of Islam were not necessarily borne out in the practice of Muslims. For example, in 1391 the black African Muslim king of Bornu (northern Nigeria) complained to the sultan of Egypt that Arab raiders were enslaving his free Muslim people, a practice forbidden by Islamic law (Lewis 1992: 53).

Arab writers were important in consolidating what was to become a widespread justi-fication for black African slavery: the 'curse of Ham'. The original Bible story tells how Ham, the son of Noah, saw his father naked and drunk. In anger, Noah cursed Ham's son Canaan to be 'a slave of slaves'. Given that the Bible story makes no mention of colour, it is interesting to enquire when and why religious commentators developed the story that Ham had been a black man and the ancestor of black peoples (with Noah's other sons linked to other peoples of the world). Goldenberg argues that, although the connection between Ham and blackness was first made in the fourth century AD and possibly earlier, the links between Ham, blackness and slavery were consolidated mainly by seventh-century Arab writers, becoming common in Islamic and Christian texts there-after. Around this time, too, Arab writers began to refer to 'blacks' as a collective category as opposed to an individual description. Again, this development was shaped by the con-text of the Muslim conquest of North Africa and more intensive encounters with black Africans, often as people being enslaved or traded as slaves (Goldenberg 2009: 156, 197).

Although Arab thinkers regarded a variety of peoples as inferior (e.g. the Turks), black Africans were often mentioned in this respect, doubtless because they, along with the Turks, were the source of many slaves. Given the importance of Islamic scholarship in Europe from the twelfth century onwards (in Renaissance movements), one might expect that Europeans would also have a negative view of black Africans. In fact, things were not so simple, because in the context of the Christian re-conquest of the Mediterranean, challenging Muslim influence, Ethiopians became seen as potential allies in Christianity. The Queen of Sheba, Caspar the King of the Moors and Saint Maurice were all represented as black Africans in religious iconography and became objects of popular veneration in many areas of late medieval Europe (Nederveen Pieterse 1992: 24–8).

The connection of black Africans with Christianity was important in fifteenth- and sixteenth-century Lisbon, where some 10,000 slaves were about 10 per cent of the city population by the 1550s. These Christian slaves were clearly distinguished from Moorish slaves, who were Muslims and seen as more recalcitrant and rebellious. Portuguese writers agreed that 'in mind and temperament all blacks were clearly adapted for servitude': they were seen as good slaves, faithful and serviceable, although sometimes supposedly given to petty theft and conceitedness (Saunders 1982: 167). Black slaves were stock characters in popular plays, and were often depicted as loyal if simple-minded servants, who might however complain behind their masters' backs. Saunders (1982: 166, 171) contends that white Portuguese saw blacks as 'innately' and 'inherently' inferior, ignorant and servile; but there is no discussion of what these terms might have meant in the context of contemporary ideas about human nature, so it is wise to take such statements with a pinch of salt, knowing already what we do about such ideas.

This late medieval linking of black Africans to Christianity was internal to Europe, reflecting European preoccupations with Islam. Encountering Africans in Africa, somewhat later, produced a rather different response, more in tune with the views of medieval Arab writers. The English did not meet sub-Saharan Africans on their home territory until after 1550, some 100 years after the Portuguese had done so, and, although small numbers of Africans had been in England from long before that, the English had mainly the classical and medieval literature as sources for theorising about them. Africans seemed very foreign to the English voyagers and traders who encountered them. Their dark skin provoked a lot of comment, and blackness was generally seen as bad and ugly – it was usually ascribed to climate and/or to the curse of Ham. Africans were seen as heathen and religiously defective, and as savage in their behaviour, including being sexually promiscuous; some resemblances were perceived between Africans and apes (Jordan 1977: 3–43).

The English tended to classify and describe Africans in a way which, by opposition, defined themselves in a positive light. In fact, at a time when the English were shifting towards ever greater mobility and military and mercantile engagement with other regions of the world, existing ideas about blackness and lightness (often called 'fairness', with connotations of both beauty and whiteness) 'rather than being mere indications of Elizabethan beauty standards or markers of moral categories, became in the early modern period the conduit through which the English began to formulate the notions of "self" and "other" so well known in Anglo-American racial discourses' (Hall 1995: 2). As with the Arabs, a context of expansion and conquest set the scene for classifying whole categories of people, perceived as other, in terms of inherited appearance.

Jordan argues that the English at this time did not have an elaborate theory about the permanence or innateness of black Africans' moral qualities. One could find occasional remarks about the 'innate' tendency to steal, a 'vicious humour' that characterised the 'whole race of Blacks' (1670) or about a 'perverse Nature' and 'all sorts of Baseness' which could not be concealed and would break out sometimes (1705), but these did not form part of a general debate about human nature (Jordan 1977: 26). Still, the similarity between these remarks and the Spanish ones cited above about 'vile race' is striking: they clearly indicate ideas about a set of connections, established at the level of collectives of people, between ancestry, nature, temperament and, in this case, appearance. There may not have been elaborate theories about innate human biology, but there clearly were ideas about engrained collective characteristics, shaped by environment and passed on through (flexible) mechanisms of heredity; these were characteristics that were seen to explain behaviour and acted as a basis for making evaluations of individuals and groups.

In further resonance with questions of ancestry, which we examined for Spain and which are, by definition, gendered, as they involve sexual relations between men and women, there were important gender dimensions to early modern English ideas of whiteness and darkness. In Spain, maintaining purity of *raza* as lineage was centrally about men controlling women in order to protect their own status and power. In the same way, the self-definition of the English as fair or white – in opposition to a subordinate blackness increasingly associated with slavery – was being carried out primarily by English men, who expected to dominate both women and black people. Blackness was typically represented by males, while 'fairness' was associated with femininity (purity, virginity, innocence, harmlessness, beauty). English women could be deemed fair and pure – and thus appropriate wives and mothers for elite men – or they could be 'blackened' and seen as inappropriate partners: men's interests were paramount here. Descriptions of blackness and whiteness could serve to talk about both 'proper' gender relations, in which fair and pure women supported the power and status of white men, and 'proper' colonial relations, in which black men and women supported the dominant position of white men (and women) (Hall 1995: 4).

2.4 New World colonisation

How did racial thinking develop in the context of European colonisation of the Americas? If conquest, and the enslavement so often associated with it, was important for the development of ideas about the character of those conquered and enslaved, and if European modernity is fundamentally linked to the conquest of other regions of the world, particularly the Americas, which became the first European colonies, in which territories and peoples were systematically and forcibly conquered, settled, exploited and controlled for material profit, then the Americas are clearly a vital context in which to examine how thinking about human diversity and hierarchy changed and in particular how racial ideas developed.

One approach is to assume that the conquest of the New World necessarily represented a 'drastic qualitative shift' in concepts of race:

> The links between religiously coded racism and colour-coded racism were the consequences of early modern European imperial expansion in Africa, the subjugation of the indigenous populations of America, and the evolution of the ancient practice of slavery into the hugely profitable transatlantic slave trade .... The history of racism as we know it today began to be articulated right then, in the sixteenth century, and there, in the Atlantic world. (Greer, Mignolo and Quilligan 2008: 2)

As we saw in Chapter 1, the coloniality approach, in which Mignolo is a key figure, links racism – 'as we know it today' – very closely to European modernity. But when we look in detail at concepts of race and the way race operated in processes of exclusion and subjugation in the early colonial world, the continuities between pre-conquest and post-conquest ideas of race in the Atlantic world are actually rather striking. While colour certainly became a key feature of race in the end, there was no sudden 'drastic' shift to colour-coded racism in the sixteenth century. Instead, a complex mixture of ideas about bodies, ancestry, morals and behaviour, all mutable and conditioned by environment (climate, food, the stars, God) continued to be dominant. These ideas now functioned in a new social context – the exploitation and control of native American populations and the more systematic use of ever larger numbers of slaves, who quickly became predominantly African – but the ideas themselves changed slowly.

Castas and *raza* in Latin America

The Spanish and Portuguese colonies had small numbers of Europeans, mainly male at first, who encountered and subjugated very varied indigenous societies. These ranged from the Inca and Aztec empires of the Andes and central Mexico, with large settled populations, where the Spanish could take advantage of existing social hierarchies and regimes of governance, to more sparsely populated regions, less easy to exploit and control (for example, in much of Brazil). In all areas the demographic decline of the indigenous population was precipitate, leading the colonists to use African slaves, above all in areas where the indigenous population was very thin and/or difficult to control; slavery was less important where indigenous labour remained viable and recovered somewhat over time.

Everywhere, interbreeding occurred in many directions. One important channel was European men and their American-born sons mixing with indigenous and African women – both through violence and coercion and through consensual unions – producing 'mestizos', who then also mixed among themselves. At the same time, the slaves were able to gain freedom on an individual basis, through 'self-purchase', thus creating a number of freed blacks who also mixed with indigenous and mestizo people. By the late eighteenth century mixed people of all kinds were a majority in many areas. Colonial society was very hierarchical, with whites at the top, slaves, freed blacks and indigenous people at the bottom, and a very heterogeneous category of mixed people – collectively called *castas* (breeds or types, defined at least notionally in terms of ancestry) or 'free people of colour' – in the middle strata.

While it sounds as if this is, in some respects at least, a 'colour-coded' system, in fact things were not straightforward, either in relation to colour or system (Cope 1994; Gotkowitz 2011; Katzew and Deans-Smith 2009; Martínez 2008; Rappaport 2014; Wade 2010: 24–30). To start with, *indio* was mainly a fiscal and juridical category, rather than a colour one (see Fig. 2.1). When the Spanish encountered indigenous people they initially assumed they were 'barbarians' – pagan, uncivilised and some of them reputed to be cannibals. They also assumed they could enslave them. Soon, however, there were doubts about the legitimacy of enslaving *indios*: there were signs of 'civilisation' in the Aztec and Inca empires; *indios* were seen not as infidels, who had rejected Christianity (as Muslims were said to have done), but as pagans who had simply not heard the word of God; as they lived in what was now Spanish territory they were also subjects of the Crown, which made it wrong to enslave them. The idea that they were 'natural slaves', in Aristotle's sense of the term, was also questionable, as this put into doubt the possibility of their redemption through religious conversion and also conflicted with the Christian ideal that all humans were the same; instead, they came to be seen as childlike humans who, in theory at least, needed protection and uplifting (Pagden 1982). Enslaving indigenous people was therefore outlawed by 1542 in the Spanish colonies and by 1570 in Brazil – although colonists nevertheless continued to enslave *indios* in a few areas, and they continued to

Figure 2.1. Amerigo [Vespucci] discovers America. Note that the indigenous woman is marked not by phenotype, but by her semi-nakedness.
© National Maritime Museum, Greenwich, London

be brutally exploited in any case. In practice, *indios* then became native people who lived in certain places – indigenous communities – and paid tribute, in money, goods or labour, to the Crown or its representatives. The category *indio* was defined by language, place of residence, clothing and occupation as much as by physical appearance or ancestry. Thus someone who was classified as *indio* could become mestizo by dressing differently, speaking Spanish and living in a town.

White (*blanco*) was a rare term until the late eighteenth century (Thomson 2011). Europeans were referred to as Spaniards (or Portuguese), even if they were locally born *criollos* (a word signifying anything locally produced). This does not mean that 'white' skin colour was insignificant – bodily appearance was still an important cue for assigning identity. In some cases colour could be the basis for collective categorisation: Spanish priests in Lima in 1648, pondering the diversity of the human species, spoke in terms of whites, browns and blacks, and conflated these with colonial categories of Spanish, *indio* and black (Silverblatt 2004: 115). But it is noteworthy that the term *blanco* was not widely used: parentage and status were key, and colour was taken for granted.

Slave was technically a straightforward legal category – either you were or you weren't – and after indigenous slavery was banned, it became virtually synonymous with *negro* (black). In contrast to *blanco*, this colour term was widely used and not taken for granted, instead marking out a subordinate class in physical terms. For the Spanish priests, above, the species-level and the colonial terms coincided for blacks, making this the most clearly colour-coded category. But even so, *negro* was not a simple matter, because the boundary between this and other categories was not clear. Differences were recognised between African-born blacks and American-born blacks (also known as *criollos*). And then there was *mestizaje*, the process of sexual, but also cultural, mixture, which produced what, by the mid-1500s, the Spanish were calling *castas*, usually used to refer to mixed people (Martínez 2008: 161–7).

The Spanish used various labels to assign people to a *casta*. A *mulato* was, in theory, the product of a union between a white and a black person. *Morisco*, used in Spain to label Muslims who had converted to Christianity and their offspring, was used in New Spain (Mexico) to refer to someone who notionally had less African ancestry than a *mulato*. Mestizo referred to a person of white–indigenous parentage. But these and other such *casta* labels were deployed in varied and inconsistent ways (see Fig. 2.2). For sixteenth- and seventeenth-century New Granada, Rappaport (2014: 5) argues that classificatory practices 'were ambiguous, ragged at the edges, overlapping, and frequently displayed no centre'. They were not part of a coherent system, but 'a series of disparate procedures that were relational in nature [and] … highly dependent upon context'. The Mexican Inquisition, investigating a woman for religious misdemeanours in the 1680s, described

Figure 2.2. A *casta* painting: 'From Spaniard and Black Woman, Mulata'.
(Museo de América, Madrid)

her as 'a white mulata with curly hair, because she is the daughter of a dark-skinned mulata and a Spaniard'; she also spoke an indigenous language, having been 'brought up among Indians' (Earle 2012: 8). One man in New Spain, being held by the Inquisition for blasphemy, was variously described as mestizo, *castizo* (of Spanish–mestizo parentage) and *mulato*; he defended his claim to be a Spaniard by identifying his parents as Spaniards and giving their occupations and places of residence (Cope 1994: 56). Individuals could be classified in different ways by administrative and religious officials, and by people in everyday usage: physical appearance might well count, as did parentage, but so did manner of dress, place of residence, occupation and comportment.

CASTA PAINTINGS

In eighteenth-century New Spain, series of paintings depicting the *castas* and their various mixtures became popular, and were displayed in both public spaces and private collections. They usually represented a man, a woman and their child, and would be labelled 'From Spaniard and Black Woman, Mulata' or 'From Spaniard and Mestiza comes Castiza'. About 100 series of paintings have been documented today. They are a formal and rather artificial representation of *mestizaje*, which almost certainly overstated the coherence of the *casta* 'system' (Katzew 2004).

There was much indeterminacy in this society of *castas*, and various factors defined status. Ideas about *limpieza de sangre* were soon adapted to the colonial context, and parentage or ancestry carried a good deal of weight, while colour functioned above all to mark out blackness. Although indeterminate in many respects, the social order as a whole was strongly structured by a classificatory hierarchy in which white/Spanish, brown/*indio* and black were the key poles, even if the middle, mixed ground was highly negotiable and flexible. Inquisitors, those 'principal arbiters of imperial culture', and other administrators were building the idea and the material practice of a unified state to govern the colonies and, in the process, 'diverse social relationships were being conjured into abstract race categories'. In seventeenth-century Peru 'magistrates and other functionaries were routinely dividing people into Spanish-Indian-black boxes as a matter of course' (Silverblatt 2004: 18).

Although the concept of *casta* depended inherently on ideas of heredity and the purity and mixing of blood and lineage, race or *raza* was not a term much used in practice: as in peninsular Spain, *raza* generally referred to Jewish or Moorish blood, but by the late 1500s people in New Spain were also claiming that their blood or lineage was free from the *raza* of blacks and *mulatos* (Martínez 2008: 163–4). Black blood, linked to slavery, was seen at this time as more damaging to purity than indigenous blood, mainly because the Spanish had originally defined natives as themselves of clean blood (not tainted by Jew or Moor). By the later seventeenth century, as concerns with defining and defending elite status multiplied – especially among *criollo* whites, whose parentage could be suspect because of their local birth – proving purity became more common and began to require freedom especially from black but also from any mixed blood. Even then, however, indigenous people themselves could still claim *limpieza*.

Flexible bodies and natures

If colour was not a straightforward matter in these racial categories, neither was fixity and immutability. The 'nature' of Spaniards, *indios* and *negros* – let alone all the intermediate *castas* – was not easy to pin down either, because human nature was not seen as a fixed thing. As we have already seen, concepts of human nature and the body at this time did not make a clear distinction between what we would now call biology and culture: a person's whole constitution was shaped by a variety of forces, including climate, food, surroundings, experience and heredity (itself seen as something that did not transmit traits in a fixed fashion).

Rebecca Earle focuses on the role food played in the Spanish colonies. She notes a shift in the historiography of this region and period: several scholars argue that racial thinking can be detected in early colonial contexts, rather than emerging only in the nineteenth century, with a focus on biologically fixed human types. These scholars claim that sixteenth-century writers already thought that 'fixed and substantial physical differences' separated Europeans from Amerindians (Earle 2012: 4). Earle, however, shows that bodies were not understood as fixed; on the contrary, they were changeable and had to be continuously maintained and cared for. This does not mean thinking was not racial – although Earle herself remains agnostic on this. It does mean we have to see race as not necessarily being only about immutable qualities; we have to grasp how people understood bodies at the time.

Bodies were thought to be shaped by 'humours' (four bodily fluids – phlegm, blood, yellow bile and black bile – which controlled individual health and temperament) and, as in theories of the ancient world, humours were influenced by climate and surroundings, including the heavens. Spaniards were seen as dominated by the 'choleric' humour (yellow bile), while the natives were mainly 'phlegmatic'. This, for example, explained the lack of facial hair among the Amerindians. The Spaniards feared that, in the colonial environment, they would lose their prized beards, which were a mark of their manhood. But food was as important as any other influence: 'As one seventeenth-century Mexican writer put it, "through eating new foods, people who come here from different climates create new blood, and this produces new humours and the new humours give rise to new abilities and conditions"' (Earle 2012: 5). By eating the right foods, dangers could be averted. The Spanish were concerned that, especially at the beginning, colonists got sick and often died. The remedy for this was to import the foods that they were used to, such as wheat bread, wine, chicken, lamb, etc., which would sustain not only beards and health, but also 'temperance and virtue' (2012: 25). The drastic decline of the indigenous population was also thought to be due partly to their consumption of new and unaccustomed foods.

Amerindian bodies were certainly different from Spanish ones, but those differences were far from fixed: 'food helped *create* the bodily differences that underpinned the European categories of Spaniard and Indian' (2012: 5). This malleability entailed a problem, however: that Amerindians could become like Spaniards by eating a Spanish diet, just as Spaniards could lose their Spanishness by failing to do so. The dilemma was at the heart of the Spanish colonial enterprise, which aimed to protect and uplift indigenous people, converting them to Christianity and 'civilising' them, but which also relied on a fundamental and enduring separation of the native population (labour, tribute payers, uncivilised) from the Spanish population (rulers, tribute receivers, civilised).

The important point is that what I would call a *racial* divide was profoundly concerned with bodies – understood as physical–moral wholes – but it did not depend either on colour alone or on ideas about fixed and immutable natures. Earle admits that it is misguided to see contemporary views of the body as infinitely flexible: blackness, for example, was seen by some as the mark of a permanent divine curse (the curse of Ham). It was also seen as a trait that was hard to shake off through interbreeding: in the rather artificial enumerations of mixings found in the *casta* paintings, while the union of a *castizo* and a Spaniard was said to produce a Spaniard, showing that indigenous blood could be left behind, the union of a *morisco* with a Spaniard produced a so-called *torna-atrás* (throw-back). Where blackness was concerned, there was less flexibility and more emphasis on colour.

Bodies were equally important to English colonists in North America, who developed ideas about the body that were quite similar to those Earle outlines for Spanish America. Bodies were seen as shaped by the environment, and this meant that English and Native American bodies were understood as being distinctive; difference was not only about levels of civilisation and savagery. Ideas of natural philosophy, such as theories about humours, underwent in the colonies a 'mutation into a racial idiom'. English bodies were seen as better adapted to the surroundings than those of Native Americans, which were susceptible to disease and thus weak. Thus 'English discourses of nature and the human body were fundamental to their imperial project. The proof was in the body' (Chaplin 1997: 232–3). As in Spanish America, colonists believed that the new environment of the Americas would produce new bodies: specifically, they talked about bodies being 'seasoned' by exposure to the American surroundings. Yet they also resisted the idea that their English constitutions would change very much: even if some English people adopted Indian habits, such as laziness, the English body remained essentially the same. Meanwhile, the idea took hold that Native American bodies were innately weak. We can see here something of a contrast with the Spanish American case – there is less flexibility and a sharper drawing of boundaries between English and Native American bodies. Joyce Chaplin (1997: 251) links this to an unwillingness to 'face the possibility of a mestizo population', which would have forced colonists to consider how heredity might work when different kinds of bodies mixed together. The English belief in their physical superiority militated against mixing with the Native Americans (although this did occur in some frontier settings, where English men were on their own). This suggests that there was a contrast between the Spanish and English colonies in terms of the frequency and role of mixture, and that this shaped ideas about race and racial difference. Let us examine this in more depth.

Gender, mixture and regional variation

We have already noted that gender and sex are integral to race: genealogical thinking, ancestry and physical appearance (understood as a shared trait, passed on by heredity) are all mediated through sexual relations between men and women. When these gender–sexual

black category (US Bureau of the Census 2002). Increasingly strict and encompassing defi-nitions of 'black' ultimately resulted in the so-called 'one-drop' rule (which became law in the twentieth century), according to which a person with any amount of black ancestry would be considered legally black.

These different regional patterns of gendered mixture were the result of various factors. The difference between a conquistador and settler colony – with the attendant differences in demographic ratios of blacks to whites and of men to women – had a major role to play. In North America the presence of a large white population, much of it plebeian, competed with and restricted the emergence of both a non-slave black population and a racially intermediate mixed population; in Latin America the conquistador whites felt less threat-ened by either of these categories of people – indeed, they could prove useful – while also being poorly placed to institute the active repression needed to prevent their emergence. The differences between English and Iberian colonists and their attitudes probably also played a role. Smedley (1993) emphasises the particularly powerful ethnocentrism of the English, who had already colonised Ireland in quite a brutal fashion, and classified the Irish as savages and barbarians. It was an easy step to apply similar modes of thought to native Americans, and especially to black Africans, who appeared in the New World as slaves. Iberian colonists, she argues, while still prejudiced against people seen as barbar-ians, had less exclusive attitudes.

It is important to be careful with such judgements about the attitudes of Iberians. It is true that they had some experience of African slavery and had legal traditions that allowed the freeing of individual slaves; they also decided to prohibit indigenous slavery. However, they were well versed in discriminating against Jews and Moors, they embraced African slavery in the colonies, and they established discriminatory and hierarchical soci-eties in the New World. But the colonial order that they created did not engender the same kind of exclusivism as in North America, where existing ethnocentrism and anti-black prejudice fed into a colonial order that quickly became based on racial segregation.

On the other hand, existing English attitudes did not determine outcomes in a rigid way. In the English colonies of the Caribbean, where whites were a very small minority – under 10% – vastly outnumbered by slaves, interracial sex between white men and black women was more common than in mainland colonies and 'mulattoes' recognised as a cat-egory separate from blacks, with freedmen (mainly 'mulattoes') constituting some 3–5% of the total in the late 1700s (Sio 1976). There was no attempt to ban interracial sex or marriage and, although marriage was uncommon, English ethnocentrism did not prevent frequent sexual relations (Jordan 1977). Thus in 1844 'coloured' (mixed-race) people were about 18% of the total Jamaican population, whereas in 1850 the United States counted a mere 1.7% of its population as 'mulatto' (Holt 1992: 215; Reuter 1918: 118).

The concept of race in the colonial Americas was shaped by these patterns of regional variation. It is often said that racial categories in Latin America today are based on colour, while in the United States they are based on descent (Nogueira 2008 [1959]). This is a crude contrast in so far as racial categories generally involve both ideas about physical appearance and ideas about ancestry and parentage. More to the point, in colonial Latin America there was an overriding concern with ancestry, because colour did not work well as a differentiator when so many people were mestizos. It did work to some extent, especially for identifying blackness, but there was a good deal of indeterminacy about racial identifications, even while race powerfully structured the social order. Racial identifications also depended to some extent on what we would now call culture: this was clearest for the identity of *indio*, which was constituted partly by language, clothing and place of residence, as well as ideas about the body.

In contrast to the Latin American colonial emphasis on ancestry, in the United States, while ancestry came ultimately to serve as the legal arbiter of race – in the context of patterns of segregation that required a workably clear definition of who was black – colour and other aspects of physical appearance actually served as the practical differentiator. Indeed, census enumerators in the nineteenth century had to rely on visual cues to determine whether someone had 'any perceptible trace of African blood'. Indians living outside their reservations were also counted after 1890, but no details were given on how to identify them: enumerators were clearly expected to be able to recognise them, whether they were living on a reservation or not (US Bureau of the Census 2002: 14, 18). This visual method worked perfectly well to maintain the strict separations between whites, blacks and Native Americans, which were a product of white insistence on sharply drawing and policing the boundary between whiteness and non-whiteness, and encompassing all whites within the privilege of whiteness, even if they were plebeian. In the United States race was a great deal less indeterminate and flexible: most people were rigidly classified into a small number of racial categories – although 'Hispanic' identity has long been ambiguously located between black and white (Gutiérrez 2009). Nature and culture might still be conflated, in that a person's moral qualities could be seen as part of their physical constitution, but categorisation was much more definite and agreed upon. (When the census started to use self-definition in 1960, the number of blacks counted barely changed: people mostly agreed on who was black.)

Race categories from below

Colonial categories such as *indio*, Indian, *negro*, black, Negro and mestizo – not to mention all the other *casta* terminology – were created by the dominant white class and imposed on the people they controlled and exploited; white was also an invention in this sense, which emerged – gradually taking over from 'Spaniard' or 'English' – to label a dominant racial group that could be quite heterogeneous. How did subordinate people so classified react to these categories?

It is clear that, over time, they accepted and used them: the forms of identification deployed by the dominant people in government and economy became the normal mode of categorisation for most people. This is not to say that subordinated people did not resist oppression. Colonial history records many slave and native rebellions, marronage (slave flight) and the establishing of maroon communities (at least in Latin America; less in North America where there was stricter control), the struggles of indigenous communities to retain their land and resist encroachment by mestizos, the petitioning of authorities by indigenous leaders, the continuance of religious practices seen by the authorities as pagan or even heretical – not to mention the simple evasion of colonial control by living in remote areas, and the more minor tactics of foot-dragging, insubordination and recalcitrance. But these practices of resistance did not often entail challenging the basic system of racialised categories in itself; in fact the category *indio* could serve precisely as a basis for claims, because it was, in theory, recognised by the Spanish authorities as entailing certain rights and protections. More common were challenges to accepting a specific category for oneself or one's community.

Some rebels did challenge the basic categories of colonial rule. Andrés Tupac Amaru, leader of the great 1780 anti-colonial insurrection in the Andes, obliged his followers to call themselves *qolla*, an Inca term. Local indigenous people used various terms, outside the colonial category system, to designate Spaniards, including *virachoca* (an Incan deity), *pukakunga* (a type of red-necked bird) and *q'ara* (a Quechua word meaning naked or bald). Most of these terms simply reproduced standard colonial categories – although *q'ara* could refer to anyone seen as an exploiter of native community members, Spanish or mestizo – but *qolla* referred to a utopian brotherhood of indigenous people, whites, mestizos and blacks, which made it a radical departure from the colonial norm. This aside, however, the insurgency tended to pit *indios* against the Spanish and whites. The insurrection itself was a highly complex process of alliances and fractures, with some *criollo* whites making tentative links to the insurgents before finally rejecting them, while some mestizos participated in popular mobilisations and some indigenous leaders fought on the side of the Crown. Yet the key axis of symbolic division emerged as *indios* versus non-*indios*, thus reinforcing dominant colonial categories (Thomson 2011).

Other Andean practices challenged Spanish rule and ideas of what *indios* should do: in the face of catechism and attempts to impose Catholic orthodoxy, local people continued to pray to the ancestors and bury their dead as they saw fit, all the while negotiating with colonial rulers for benefits for their communities. So-called 'Andean virgins' combined images of Christian nuns with Incan ideas about virginal servants of the sun god and promoted nativist movements that aimed to re-establish Inca ways of life in the Andes (Silverblatt 1994). The virginal status of these women gave them social legitimacy and protected them to some extent from the interference of colonial authority. Such practices were 'fashioned in, but were not reducible to, the antagonisms of colonial life': they displayed some autonomy, but within the confines of the dominant world; they were 'ultimately bound by colonial terms of living' (Silverblatt 2004: 211).

This suggests that attempting to renegotiate position within the existing system of rule and classification was the commonest approach. For example, in New Granada free people of colour were nearly half the population by the late eighteenth century, and many lived at the margins of urban colonial authority. Local authorities saw such people as restless, troublesome, immoral and generally lacking in 'honour', a complex notion connoting status and virtue, defined by ancestry, wealth, character (good manners and temperament) and moral qualities (honesty, obedience, respect). The free people of colour resisted such imputations and attempts by the authorities to instil greater control: especially they complained that assumptions were being made on the basis of their colour and ancestry, rather than their actual behaviour, which displayed the requisite honourable qualities.

A militia captain *pardo* (brown person, of African ancestry), for example, protested that his legitimate authority was being undermined by the local judge and the mayor, who prompted his subordinates to disobey him, refused to allow him to discipline them and then accused him of insubordination and sent him to prison. He complained that they were discriminating against him because of his colour and ancestry, when his service record showed he was an honourable man. This was a case of colonial authorities resisting what they saw as the encroachment of free people of colour into domains of authority that they had controlled hitherto (Garrido 2005). The challenge mounted by these people was complex: they were denying that judgements about character could be made on the basis of colour and ancestry; honour was not inherited, but earned. This contested some elements of the racial order. On the other hand, the basic set of racial categories remained in place and the key value of honour as an organising axis for the system was reinforced (Johnson and Lipsett-Rivera 1998).

In North America the weight of colonial rule was stricter still, especially as slavery became embedded. In seventeenth-century Virginia freed blacks could enjoy some autonomy. Philip Mongon, for example, had sexual relations with white women (he was accused of adultery, as he was already married) and, when charged with stealing hogs in 1659, defied the judges' authority by throwing pig's ears on the courtroom table; he borrowed money from whites and socialised with them as friends and neighbours. Other such free black families owned land and slaves, and not only borrowed from but lent to whites. Freed blacks tended to assimilate into colonial English society; while Mongon retained an African surname, in most respects he fitted into Virginia society and, like other free black families, distanced himself from slaves. By the late 1600s, however, freedom was more difficult to obtain, strictures on freed people increased markedly – for example, freed blacks had to leave Virginia – and interracial sexual relations were banned. Free blacks could no longer distance themselves so easily from slaves: colour defined status (Morgan 1998).

Given the inherent interweaving of race and gender, it is to be expected that some of the negotiation of racial categories took place through the medium of sexual relations. Again, there were powerful limits to the room for manoeuvre. In eighteenth-century North Carolina it was not unusual for English colonists to have sexual relations with native American women, ranging from casual sex to long-term relationships and marriage (Fischer 2002: 55–97). This was thought to smooth trade relations but also to be a mode by which native women could be acculturated and, in effect, colonised, bringing Native American land into the English property system. Native Americans (men as well as women) also saw the liaisons in quite an instrumental way, but they set the terms of the exchange between English men and native women in ways that met their own interests. As the century wore on, and the territory became increasingly colonised, with war, disease and alcohol taking their toll, Native American autonomy was reduced and English men took the upper hand. In addition, English tolerance of interracial marriages became less and ideas about race hardened, so that native and English bodies were increasingly seen as different.

Some lower-class white women in North Carolina also challenged racial and sexual boundaries by having relationships, and children, with black men. Again, however, the potential of such unruly activities to unsettle the colonial social and racial order was constrained, and the authorities moved to ban interracial marriages and forced children of illicit unions to be indentured servants for thirty-one years. White women could also contribute to the hardening of racial boundaries in their reactions to slander: in court they defended themselves against accusations that they had had sex with black men as this impugned their honour (Fischer 2002: 131–58).

CONCLUSION

This chapter has covered a lot of ground, from the ancient world to the early modern American colonies of European powers. The aim has been to trace emerging racial thought, as it has been shaped by the context of European colonialism and modernity. Ancient and medieval ways of thinking about human diversity, 'others' and especially specific categories of others (Jews, Moors, African blacks) set the stage for the development of racial thinking in the New World. That stage was already provided with a toolkit of concepts that allowed people to locate human differences, perceived as shared, rather than just individual, in the domain of nature, where nature was defined as encompassing climate, the geographical environment and its humeral influences, the heavens and divine powers, as well as aspects of the human constitution, not split easily into what we might now call biology and culture, or nature and nurture, or body and mind. Such human differences were also linked to patterns of heredity and blood, although not inflexibly, as neither of these things was seen as entailing permanency. More generally, nature was not seen as something fixed; that something was 'natural' did not necessarily mean it was immutable, as we might be tempted to assume today (mysteriously, when one thinks about it, as nature is clearly a domain in constant flux). The toolkit of concepts dealing with difference, blood, bodies and environments – plus ideas about Jews, Moors and Africans in particular – certainly has close affinities with racial thinking, but, as I have already argued, I prefer to retain the concept of race for the way these concepts developed in the context of European colonialism.

Within this context, I have advanced two further arguments. The first is that it is misleading to characterise thinking about human difference in the early period of the colonies as non-racial or as a kind of 'cultural' or religious mode of racial thought on the grounds either (a) that people were only concerned with a 'classification of cultures' (Todorov 1993: 101) or religious differences and were not concerned with bodies; or (b) that people were not working with a theory of 'biological' difference, understood as entailing permanent differences passed on immutably through heredity. In relation to (a), I think it is abundantly clear by now that people were certainly concerned with bodies as a locus of human difference. With regard to (b), if racial thinking is defined only in terms of theories about 'supposedly permanent and indelible bodily traits, whose significance is located precisely in their immutability', then it is indeed 'difficult to speak of an early modern idea of race' (Earle 2012: 188). Or 'if race in modern times signifies a fixed set of bodily traits, purportedly specific to national or ethnic groups and transmitted through procreation' then admittedly 'it was not a coherent hypothesis in the early modern period' (Chaplin 1997: 230). But I think it is too narrow to define racial thinking in this way. We need to expand our purview to include the flexible ways of thinking about nature, bodies and minds that we have observed in this chapter: race thinking is not *only* about fixing, it can also be about mutable human constitutions.

The second argument is that the advent of colonialism did not entail a drastic and sudden shift from ideas about religion and culture to concepts of race simply in terms of colour. Skin colour and other somatic features were indeed made into important markers of difference between the populations – principally Africans, native Americans and Europeans – that New World colonialism brought into relations of domination and subordination, but ideas of ancestry, filiation and blood were also vital and continued to be so. Particularly for relations involving Europeans or whites more generally, indigenous people and mestizos, skin colour was only a partial means of identification. For relations involving Africans and their descendants, skin colour assumed greater importance and acted as a distinctive marker. But ancestry and filiation were still crucial: black blood was seen, in the social order of the *castas*, to be particularly hard to shake off, genealogically speaking; it tended to make itself visible. This indicates that ancestry and physical appearance – the hidden inside and the visible outside – are closely linked in racial thinking: hence the assumption that the US census enumerators could enforce the 'one-drop' rule of ancestry simply by looking at people.

A further conclusion derived from the material in this chapter is the importance of conquest as a context for elaborating ideas about others seen as subordinate and inferior; and especially conquest that results in enslavement. But the New World colonies show us that the exact circumstances of conquest and the subsequent social order shape the development of racial concepts. The differences in the overall structure and functioning of the English and Spanish colonies were key to the emergence in the former of a more racially segregated society, in which skin colour and ancestry together came to define status in relatively inflexible ways. In the Spanish and Portuguese colonies there emerged social orders that were varied but that generally gave a very different place to mixed people, creating less rigidity in the definition of social status and location, although this depended to some extent on where a person was in the social order: for black people (and indeed for white people) colour tended to define status more inflexibly. Again, the presence of more and less flexibility here does not make the English colonies 'more racial' than the others, as if race were always defined by inflexibility: this kind of contrastive comparison is tempting, but should be avoided. All these colonies – and I have only mentioned the Caribbean islands in passing and not dealt at all with French Canada – are variations on the same basic themes of conquest, slavery and settlement; they are all variations on the same theme of race thinking.

Finally, we have seen that gender relations are integral to the way racial thinking works and how racial relations are constituted. Policing the boundaries of a group, with a view to defining the subjects of privilege and the objects of exploitation, always involves policing the sexual boundaries of the group too. In so far as men consider themselves, rather than women, to be the ones who define privilege, then women's sexuality will become an object both in the policing of boundaries and in the business of exploitation. This was common all over the Americas, but it worked itself out very differently according to the regional variations noted above. In the Spanish and Portuguese colonies men's sexual relations with non-white women were not only more frequent than in the mainland English colonies, but were allowed to result (or could not really be prevented from resulting) in large numbers of people socially recognised as intermediate between whites, blacks and indigenous people. This underlay the emergence of a distinctive way of thinking about race, which displayed an obsession with ancestry and also, especially for the indigenous–mestizo boundary, a concern with behaviour as a defining trait of race. Clearly such a system was not very amenable to rigid segregations.

FURTHER RESEARCH

1. The books by Snowden (1983) and Thompson (1989) have many images of black people in the ancient world; you can find images online too. From these images, what would you conclude about Graeco-Roman views of Africans? Find some images of 'slavery in colonial Virginia' and compare them. What does this tell you about changing views of race?

2. *Casta* paintings. There are lots of online resources to explore these depictions (e.g. Pooley 2010). Try to investigate how these images relate to the context of eighteenth-century New Spain. What do they tell us about 'race' at the time?

3 From Enlightenment to eugenics

The period we explore in this chapter is a long and varied one: from the Enlightenment (dating from about 1650 to about 1800) to the era of eugenics (from the 1880s to the 1930s). The last chapter, although it started with the ancient and medieval periods, also explored early modern New World contexts, reaching into the late eighteenth century, and thus overlaps considerably with this chapter. But I want to keep this very broad period in mind, because it helps to grasp a series of transitions – but also continuities – which are vital to understanding changing ideas about race and changing structures of racialised relations.

As we saw in the previous chapter, some scholars speak of a transition from a pre-1800 classification of cultures to a post-1800 classification of bodies. I think this is misleading in so far as it ignores the role of ideas about bodies, nature, heredity and blood in earlier periods – although these ideas need to be understood in their historical context. It is also potentially misleading by inviting us to focus only on bodies (or more specifically biology) after 1800, neglecting the role of ideas about culture and behaviour, or rather neglecting how these might have been understood to shape biology, rather than biology determining culture in a one-way relation. As I have argued, race has always to be seen as a natural–cultural assemblage in which 'nature' and 'culture' are always shaping each other and the differences between them are not always clear.

However, there is no doubt that important changes in racial thinking did take place, especially during the nineteenth century, although these have to be seen in the context of the longer period spanned by this chapter. By 1850 the Scottish medic and anatomist Robert Knox could say, 'That race is everything, is simply a fact, the most remarkable, the most comprehensive, which philosophy has ever announced. Race is everything: literature, science, art – in a word, civilization depends on it' (Knox 1850). This statement would not have been made in 1650, or even 1750. As I have argued in previous chapters, the absence of a declaration of this kind in these earlier periods should not be taken to mean that racial thinking did not exist then: it was not organised under the banner of 'race' as a coherent and elaborated theory (about biology), but it certainly existed.

Knox's statement was linked to two major shifts. First, the idea of race, explicitly named as such, became central as a way to explain a whole range of human phenomena: origins, behaviour, diversity, social organisation, intellectual production, moral value and place in a global hierarchy. Second, physiological variation, to be assessed through scientific techniques of anatomy and biology, became the basis to define certain types (or races) of humans, whose moral, behavioural and intellectual qualities were closely interwoven with that physiology – although, as noted above, it was not a simple one-way determination.

3.1 Transitions

How are we to understand the changes that led to Knox's statement becoming possible? Reviewing a series of 'transitions' that took place during this period – with the vital qualification that each transition also involved important continuities – will give us the context for understanding changing ideas about race.

Science

Explaining the diversity of human character and constitution has always been part of general theories about the world, the cosmos and its living beings, theories that, in the West, were strongly linked to Christian theology. From the sixteenth century in Europe some thinkers – famous names include Copernicus, Galileo, René Descartes, Thomas Hobbes, John Locke, Francis Bacon and Isaac Newton – began to look away from theological texts as sources of knowledge and more towards controlled and systematic empirical observation in order to describe the natural world in an objective way and deduce explanations for natural phenomena. (Europeans were not alone or, arguably, first in this respect: maths, astronomy and other sciences had important precedents and parallels in Islamic areas and in India and China too.) The natural world was increasingly envisaged as machine-like – operating according to regular principles, which could be revealed through careful experiment and measurement. Although sometimes called a scientific 'revolution', this change was long drawn out and by no means overturned the authority of theological knowledge, but rather incorporated ideas about the ultimate role of divine causes. The eighteenth-century Swedish naturalist Linnaeus saw himself as revealing both truth and God's plan by classifying living things into an orderly system, the principles of which are still in use today.

Science was based on the idea of an innate human rationality, which, combined with the empirical observation of phenomena, could deduce causes and effects in the workings of nature, with an eye to being able to control such workings. While sight was a vital sense, not everything was easily visible; the naked eye might need assistance (e.g. the telescope, the microscope) and nature needed to be interrogated into giving up 'her' secrets (Merchant 2006). Science became linked with truth: a Dutch philosopher defined science as 'either taken largely to signifie any cognition or true assent; or, strictly, a firm and infallible one; or, lastly, an assent of propositions made known by the cause and effect' (Franco Burgersdicius' *Logic*, II. xx. 99; 1697 translation of 1626 original). Discovering truth took a systematic, laborious and concerted programme of work.

Scientific advance was welcomed by economic elites and entrepreneurs, who were also using reason and the techniques of systematic observation, control and measurement in the practice of reckoning and accounting, with the aim of accumulating wealth. Key texts and tools in financial accounting were developed in Italy in the late fifteenth century:

double-entry bookkeeping, based on the memorandum, the journal and the ledger, were explained in Pacioli's text, *Details of Calculation and Recording* (1494). In early eighteenth-century Holland merchants and skilled craftsmen commissioned lectures on Newtonian optics, while in England the Royal Society aimed to create a more rational social order through scientific advance; Puritans also supported Baconian science as a tool for social reforms in health and education. Scientific shifts towards the understanding and control of the mechanisms of nature, far from standing alone as a theoretical endeavour, were embedded in processes of economic and political change, which were being guided towards profit and improvement and, more generally, towards the understanding and control of money and society through 'calculation and recording' (Jacob 1988).

Science was developing new ways of thinking about the natural world. First, as noted above, nature was increasingly being seen as machine-like and the role given to God in organising nature's mechanisms was steadily decreasing. Second, long-standing ideas of a Great Chain of Being, current since the ancient Greeks, were still respected in the late eighteenth century, but would begin to lose authority. The Great Chain was a static concept that arranged all living beings in a single unitary scale, designed by God, with the simplest forms at the bottom, angels and God at the top, and humans in the middle. Such ideas would be increasingly challenged by the work of classifiers such as Linnaeus and John Ray, whose classification systems did not match the pattern of a single chain. Third, the static quality of the Great Chain was undergoing revision, taking on a more developmentalist dimension that envisaged more complex forms emerging, over time, out of simpler ones; there was also the idea that human society had developed over time, from simple early forms, based on hunting and pastoralism, to later more complex civilisations. By the nineteenth century such evolutionary approaches were increasingly dominant and were based on biological processes, although it was not until the later 1800s that Darwinian explanations for how evolution worked were adopted. Finally, and most significant for our purposes, the development of science opened new possibilities for thinking about humans as organic entities, whose true workings could be explained in terms of natural biological mechanisms, observable with new technologies such as the microscope. This laid the basis for the biological science of race of the nineteenth century.

Slavery and abolition

Over this period slavery both reached its apogee in the Americas and was also gradually dismantled as a legal institution. (Informal exploitation amounting in practice to slavery continues to this day.) Figures for the African slave trade to the Americas are not definitive, but Table 3.1 gives some idea of the variation over time and place (see also Curtin 1969; Eltis and Richardson 2010).

Notable here is that the British territories that would become the United States do not appear in the table as a main destination: the area accounted for only about 5 per cent of the total number of African slaves imported to the Americas. But this does not give a true picture of the importance of slavery and the slave population in the United States. In the Caribbean and tropical South America life expectancy was low for slaves and free

TABLE 3.1 Overall share of slave trade by period and major destination areas

Period	Numbers	Principal destination areas
1500–1600	199,000	(i) Spanish American mines and agriculture (ii) NE Brazil sugar plantations
1600–1700	1,523,000	(i) Brazilian sugar (ii) Caribbean islands (pre-1650, tobacco; post-1650, sugar) (iii) Spanish America
1700–1800	5,610,000	(i) Caribbean sugar (especially Jamaica and St Domingue [Haiti]) (ii) Brazilian sugar
1800–66	3,371,000	(i) Brazilian sugar and coffee (ii) Cuban sugar

Source: http://www.slavevoyages.org

people alike, and slave numbers had to be maintained through constant importation. In the United States life expectancy was longer for both free people and slaves, and the slave population grew naturally, especially when the British restricted the slave trade after 1810 (Blackburn 1998). By the 1820s, despite the small number of imports, the United States had about one-third of all the slaves in the Americas, roughly the same proportion as Brazil, which took more than a third of the total trade. By 1860, when slavery had been abolished in many areas, with the notable exception of both the United States and Brazil (and Cuba), the US South was home to about two-thirds of New World slaves.

The important point to take from these data is the massive increase in the number of black slaves in particular areas of the New World after 1700: the English North American colonies, Brazil and the main Caribbean island colonies (Jamaica, Cuba, Haiti). In the rest of Spanish America slavery was becoming less important by the late eighteenth century and slaves formed only about 4 per cent of the population, apart from Cuba where they were about one-third of the total. Where slavery was important it was driven by demand for key commodities that were part of a global mercantile and industrialising economy: primarily sugar and cotton, with coffee and tobacco occupying a lesser role. Sugar, in particular, became an essential commodity for the urban working classes in Europe and North America: in Britain consumption increased twenty-five-fold between 1600 and 1850, while the quantity of sugar on the world market more than doubled between 1800 and 1830, reaching more than 572,000 tons; a hundred years later it was 30 million tons (Mintz 1985).

After 1800 slavery continued to be central to the United States, Brazil and Cuba, and remained legal there for a considerable time. While indigenous slavery had been outlawed in the Americas in the sixteenth century, the institution of slavery itself and the enslavement of Africans was not seriously questioned until the late 1700s. The first target was the slave trade itself, which was banned first by Denmark in 1802 and by Britain in 1807; by 1830 all other European countries had followed suit, although the trade continued illegally into the 1860s, mainly to Brazil and Cuba. The abolition of slavery itself was slower and more uneven (see Table 3.2):

TABLE 3.2 Abolition of slavery, dates by country

Country	'Free womb' laws*	Abolition
Chile	1823	
Central America	1824	
Mexico	1830	
Colombia	1821	1851
Bolivia	1831	1851
Ecuador	1821	1852
Argentina	1813	1853
Venezuela	1821	1854
Peru	1821	1854
Puerto Rico	1870	1873
Cuba	1870	1886
Brazil	1871	1888
Britain		1833
France		1848
Denmark		1848
Holland		1863
USA		1862

Note: Free womb laws freed the offspring born to slave mothers.

The causes of abolition are complex, and have been the subject of debate about the relative weight of moral and economic factors. It is fair to say that many deemed slavery morally wrong in societies supposedly based on liberal principles of liberty and equality, while it also began to seem unsuitable for increasingly capitalist economies, based on 'free' wage labour (i.e. workers who were obliged to offer their labour for sale on an open market). None of this stopped slavery continuing in highly capitalist economic sectors (such as the sugar, cotton and coffee plantations of the US South, Cuba and Brazil), nor in societies such as the United States, home to revolutionary principles of liberty and equality. But the combination of moral and economic forces gradually gained ground – albeit via a bloody civil war in the United States (Eltis 1987; Tomich 2004). Slavery was replaced by wage labour, but also by semi-free forms such as indentured labour. The latter involved about a million workers from Asia (mainly India and China) being shipped principally to the Caribbean (although further numbers went to East Africa and Mauritius), where they worked in conditions similar to slavery, often on sugar plantations, in order to pay off their indenture (the money that had been loaned to get them to their destination).

There is no simple relation between changes in racial thinking and the uneven, protracted transition from slavery to abolition, as if ideas of black biological inferiority simply replaced slavery as a tool of domination. Being in favour of abolition – for example, on religious or humanitarian grounds – did not necessarily equate to thinking that black people were equal to white people; on the contrary, explicit theories suggesting they were not equal could act as a means to justify their subordination, even in the absence of legal slavery. On the other hand, one might – and some US slave-holders did – oppose scientific theories about blacks and whites as very different sorts of humans (different species, for example), because this threatened Christian ideas about the unity of humans before God (Banton 1987: 9, 45). In addition, scientific theories about the inferiority of non-whites did not simply 'take over' from slavery as an instrument of oppression. Slavery only defined the status of some oppressed non-white peoples; and it was only in the southern United States, and perhaps in some British Caribbean islands, that 'slave' was more or less coterminous with 'black' – elsewhere in the Americas many black people were free. In that sense, racial theories were a lot more comprehensive and encompassed much more than the category 'slave'.

Still, it is hard to avoid the conclusion that there was some connection, albeit indirect, between the demise of slavery as an accepted institution and the emergence of detailed and explicit theories about the races of mankind and their relative positions in a natural hierarchy. It was '"at the very point in time when large numbers of men and women were beginning to question the moral legitimacy of slavery" that the idea of race came into its own' (Fields 1982: 153, citing Christopher Lasch). The 'intensity of attention' given in the United States to the place of 'the Negro' in the natural order was 'a besieged culture's response to the rise of militant abolitionism, the threat of emancipation and white fear of irrevocable social changes' (Smedley 1993: 236). Any connection between abolition and racial theory, however, was clearly mediated by the changing context of colonialism and imperialism.

Colonies, empires

From previous chapters, it is clear that conquest and colonialism are important contexts for grasping the role of ideas about human diversity and, especially, ideas that colonisers have of the people they are conquering, enslaving and colonising. For the period covered in this chapter there were a number of important and in some ways divergent processes at work, which are important as background for understanding changes in racial thinking. First, most of the old New World colonies became independent, mainly during the nineteenth century:

* Spanish colonies won independence between the 1810s and the 1830s (except Cuba and Puerto Rico, 1898).
* Brazil became the centre of the Portuguese empire (1822–89), before becoming a republic.

- French colonies: Haiti won independence in 1804; other French possessions became independent during the 1900s or still belong to France (e.g. Martinique, Guadeloupe, Guyane).
- British colonies: the USA declared independence in 1776; Canada was effectively independent from 1867; British possessions in the Caribbean became independent during the 1900s.

The nineteenth-century independent republics generally adopted political regimes based on liberal ideas of equality, even as elites governed societies where social and racial hierarchies were very marked and, in many cases, slavery still existed. I will explore these contradictions and their implications for the development of racial theory below.

Second, European powers acquired new colonies in a second wave of imperialism. Starting in the late 1700s, and accelerating rapidly in the mid- to late 1800s, vast swathes of South and South East Asia (and some parts of East Asia), Oceania (including Australia) and Africa were brought under the direct control of European powers. In the late 1800s Russia controlled a huge territory from Eastern Europe, through Central Asia to the Far East, where it met the Japanese empire, which expanded from about 1868 into China, Korea and beyond. By the early 1900s over half the world's land mass and nearly half of the global population were under European colonial domination; in the 1920s the British Empire alone controlled about a fifth of the world's territory and population. Just between about 1870 and 1900, the 'scramble for Africa' resulted in almost all of the continent – about one-fifth of the world's land area – coming under European imperial control.

The United States played a different role in this wave of imperialism. Rather than ruling over many colonies, it took a large swathe of northern Mexico after the Mexican–American war (1846–8), acquired Hawaii, Puerto Rico, Guam and the Philippines in 1898 as US possessions – the last three after the Spanish–American war – and occupied various countries (e.g. Cuba, Dominican Republic, Haiti, Nicaragua) for short periods during the 1900s, as part of the powerful influence the US government exerted over its 'back yard'.

Imperialism was driven in part by competition for dominance between European powers, but the underlying force was the growth of a globalising, industrialising economy led by the United States and European countries, which needed raw materials, new commodities (such as sugar, as we saw above) and new markets. Such needs did not necessarily have to be satisfied by imperial forms of control, which could be expensive and burdensome to administer; and the British, for example, also traded and invested outside their colonies. In that sense, empire expressed a complex combination of political, economic and indeed symbolic drivers. In addition, as with abolition, the connections between imperialism and new styles of racial thinking are sometimes indirect – it is not easy to tie specific scientific theories to specific events in an imperial timeline. The key point is that these empires generally involved people who identified themselves as white controlling others whom they saw as non-white. (In the Japanese case the racial dimensions of imperial rule were not so clearly colour-coded – on which more below.) This clearly created, or rather expanded, a context in which racial thinking could fulfil important functions.

Nationalism

If imperialism was linked to the jockeying between European powers, this was also because the period under discussion here witnessed the strengthening of nationalism, or even its birth. Nation is an old term referring to a people, who share a common origin and customs, but the idea of a nation as an organised political entity – what might previously have been called a realm, kingdom or country – associated with a defined territory over which the nation's people assumed a right to sovereign government really emerged in the late 1700s, linked in part to colonial territories in the Americas gearing up to throw off the shackles of metropolitan rule. Nation in this new sense was inherently linked to nationalism, a way of thinking about the nation and its people, which involved 'imagining' oneself as sharing experiences and habits, in space and in past and future time, with other members of the nation. These were people beyond the confines of family and local community, whom one did not know and would never know personally, but with whom, precisely because they were members of the same nation, one felt some affinity and shared belonging, and a common sense of ownership of the national territory and polity (Anderson 1983).

Nationalism developed strongly in nineteenth-century Europe, as nation-states defined themselves (e.g. Italy and Germany, which completed unification in the 1870s) and competed politically and economically. Governing elites were concerned with forming and managing national populations, who identified with the nation and had the mental and physical capacities to make it grow and prosper. Nationalism involved powerful ideas about common origins and could easily adopt idioms of shared blood, kinship and genealogy, as well as shared language, history and spirit, soul or culture (Alonso 1994). Place of birth and blood are often seen as opposite ways of defining membership of the nation, in legal and political terms: membership by national location of birth (*jus soli*, right of the soil) versus membership by nationality of parents (*jus sanguinis*, right of blood). But as well as providing a shared milieu, the nation's soil has often been seen as creating shared bodily substance or essence – as we saw in Chapter 2, food and environment may be thought to shape bodies and blood – meaning that birthplace and blood become interwoven (Linke 1999; Porqueres i Gené 2007).

In sum, nations could often be conceived in racialised terms, being defined in terms of bodies as well as histories and cultures. In so far as imperialism was an expression of national pride and pretension, the domination of non-white peoples could mesh easily with ideas about the nation's people as sharing blood and substance, expressed in bodily appearance. Needless to say, the notion that one's own nation was superior to others also worked easily with an assumption of superiority over non-white peoples.

Liberalism and inequality

New republics and established nations alike, in Europe and the Americas, were shaped by the tenets of liberal political thought, which emerged with eighteenth-century Enlightenment thinkers, challenging monarchy and religion as the basis of government and forming the philosophical foundation of modern democratic systems. Key principles of liberalism were enshrined in the 1776 US Declaration of Independence, which stated 'that all men are created equal, that they are endowed by their creator with certain unalienable rights, that among these are life, liberty, and the pursuit of happiness', and in the French Revolution's Declaration of the Rights of Man and the Citizen (1789), which declared: 'Men are born and remain free and equal in rights' and defined the 'natural and imprescriptible rights of man' as being 'liberty, property, security, and resistance to oppression'. What the Declaration calls 'social distinctions' may exist, but they should be 'founded only upon the general good'. It is worth noting that the Declaration did not outlaw slavery, despite lobbying on this front from the abolitionist group Les Amis des Noirs (The Friends of the Blacks). One can only conclude that slavery was deemed by the revolutionaries to be for 'the general good'.

A theoretical problem inherent to liberalism is the difficulty of ensuring that some people's pursuit of liberty and happiness does not constrain the opportunities of others, especially when some people start in a position where they have more liberty, property and security than others and thus have an advantage. Also problematic is the assumption that everyone will share a view on what constitutes liberty, security, etc. The Declaration states that 'liberty consists in the freedom to do everything which injures no one else', but people may have different ideas about what constitutes an injury: a 'harmless joke' may be taken as an insult; occupying an apparently 'ownerless' piece of land in Australia may offend against the spiritual relationship that an Aboriginal dweller has with the land and unseen ancestors and spirit beings. Equality is not easy to legislate for.

The outward manifestation of these difficulties is easily seen in the fact that liberal thought coexisted from the beginning with very substantial inequalities, accepted as perfectly normal within liberal political orders and apparently as contributing to 'the general good'. As noted in Chapter 1, in the Americas racist immigration policies were normal for much of the nineteenth and twentieth centuries, while in many countries women, the poor and the illiterate – with the last two categories including many non-white people – were often excluded from voting rights, sometimes until well into the twentieth century (Engerman and Sokoloff 2005; FitzGerald and Cook-Martín 2014). In some cases men had to be property owners to vote, which is difficult to square with all men being born with an equal right to property. The justification for inequality – which of course was not only in terms of voting rights, but encompassed huge inequalities of wealth and power and the persistence of legal slavery – was made in terms of the relative capacity of different sets of people to exercise their rights and discharge their obligations. John Locke, one of the seventeenth-century founders of liberal thought, insisted on the need to teach the reason that, at another level, he saw as an innate capacity of all humans: children had to have a proper tutor – admitted to be an expensive acquisition – who was 'well-bred' and understood 'the measures of civility'. The full possession of the cultivated reason that was needed to participate properly in the political order, and above all to rule over others, was thus only accessible to a certain class of people (Mehta 1997: 69).

A later nineteenth-century theorist of liberalism, John Stuart Mill, also answered the key question about 'who should rule' in a liberal political order by drawing divides between those qualified to rule and others who were less so. In his view 'despotism is a legitimate mode of government in dealing with barbarians, provided the end be their improvement'. Until people had 'become capable of being improved by free and equal discussion', they should practise 'implicit obedience' to a ruler (Mill 1859: 19). Mill worked in the colonial service in India, a place that in his view required direction by a more civilised ruling power (Goldberg 1993: 35). The Chinese too had 'become stationary' and if they were 'ever to be farther improved, it must be by foreigners', such as European nations, which were 'an improving, instead of a stationary portion of mankind' (Mill 1859: 135).

Liberalism thus squared the circle of equality and hierarchy by positing differences of condition between groups of people – women versus men, educated versus uneducated, civilised versus uncivilised, stationary versus improving – which legitimised the rule of one set (typically educated, well-bred, European, white men) over others (typically women, children, the uneducated and the non-white colonised populations). These differences of condition did not have to be natural – Mill, for example, contested the idea that women were naturally inferior to men and pointed out that hierarchical rule was often falsely legitimised in terms of nature – but it is obvious that theories claiming to show scientifically that some classes of people were naturally and, more specifically, biologically inferior could act as a powerful legitimisation of hierarchy in a liberal order based on ideals of equality.

Blood and sexuality

One way of thinking about the transitions in process, which also brings us closer to the character of racial thinking over this period, is via Michel Foucault's influential analysis of a gradual and always incomplete shift from what he calls a 'symbolics of blood' to an 'analytics of sexuality', beginning in the seventeenth century and being consolidated in the nineteenth (Foucault 1998). Pre-modern forms of governance depended on the power of law wielded by a monarch, who held direct control over things, people and time in his or her realm; the sovereign had the power and the right to dispose of these things, including by decreeing death. Death ordained in this way was typically by public execution, often involving torture, which displayed the power of the sovereign over the life and death of his subjects' physical bodies and blood. Social life was regulated by 'the deployment of alliance', in which marital relations between families – connections of blood and genealogies – were governed by legal and religious codes that sought to maintain a secure and stable social and moral order.

Gradually, as economic and political structures changed – in the kind of transitions I have described above – an analytics of sexuality became added to, rather than displacing, this symbolics of blood, which could no longer effectively govern a society being transformed by population growth, industrialisation and urbanisation. The exercise of power increasingly operated through bio-power, which focused on sexuality and, more generally, the active production of life, rather than the right to decree death. Governance became a matter of producing, optimising and regulating the life force and vitality of society, which was embodied in the sexuality of its members. Life, or life force, became a matter to be administered and managed in order to promote vital and vigorous citizens, nations, races and, ultimately, the human species. Expansion and optimisation became the aim, rather than the simple maintenance of stability and the status quo. Sex became a matter of intense concern because 'sex is the means of access both to the life of the body and the life of the species' (Foucault 1998: 146).

This movement between body and species is important, because Foucault was interested in how bio-power worked both by disciplining the life of the individual body, defining normal and deviant behaviours and inculcating habits seen as proper and correct and by regulating the life of the population or species: 'The disciplines of the body and the regulations of the population constituted the two poles around which the organization of power over life was deployed' (1998: 139). Science developed specialisms that interrogated life at the level of the body and the population – to understand sexuality, sanity, criminality, deviancy – and did so increasingly in terms of biological processes. This, of course, is crucial to grasping the increasing emphasis on biology as a key aspect of racial thinking: this emphasis was not just a matter of racial and scientific theories, but was also woven into a more pervasive concern with the role of 'life' in the management of the political order.

However, the symbolics of blood was not supplanted by the deployment of sexuality; instead, the latter was superimposed on the former: 'the preoccupation with blood and the law has for nearly two centuries haunted the administration of sexuality' (Foucault 1998: 149). This was linked directly to ideas about race in so far as, in the nineteenth century, according to Foucault, ideas about regulating and optimising the life of populations and bodies 'received their colour and their justification from the mythical concern with protecting the purity of the blood and ensuring the triumph of the race' (1998: 149). A preoccupation with life and biology in general meshed with attention to the life and biology of particular sets of people seen as distinctive and linked by descent: races, nations. Racism in this view is 'primarily a way of introducing a break into the domain of life that is under power's control: the break between what must live and what must die' (Foucault 2003: 254) or, in slightly less dramatic terms, between those to be included and those to be excluded from certain rights and privileges.

Racism, in Foucault's analysis, was oddly distant from colonialism. He locates early uses of race in seventeenth-century European discourses about 'race wars' between groups defined by language and habits – Normans, Saxons, Franks – who were engaged in local power struggles; the European nobility was also concerned with maintaining purity of blood and fending off encroachment by internal enemies. In the eighteenth century race became defined in more anatomical and morphological terms, with the global dominance of one race over another seen increasingly as part of the natural order. By the nineteenth century this had developed into a state-driven, biological racism that defined certain categories of people as inferior and subject to exclusion; these people were usually internal to European societies, such as Jews, but the lower classes might also be defined as being in danger of racial degeneration and a threat to the vitality and moral order of the nation. Long-standing ideas about blood could operate at the populational level of 'life', as well as the family level of genealogy (Stoler 1995: chs. 2 and 3; Taylor 2011).

The role of colonialism and imperialism in Foucault's analysis is very muted, which is frankly problematic, as it seems impossible to understand racial discourse without sustained attention to slavery and colonialism (Stoler 1995), but his approach is particularly useful in illuminating the growing role of biological thinking in science and politics, and the way in which older notions of blood are not displaced but take on new meanings. It is also useful because it gives us an excellent vantage point to see again how race and gender articulate together, via sex. We saw in Chapter 2 that race and gender articulated with each other, via sex, through notions of honour, genealogy and purity of blood. Foucault's emphasis on sexuality as a key concern of bio-power renews the idea that a concern with race was always also a concern with the sexual behaviour of gendered individuals, as both things involved the management of 'life'.

3.2 Changing racial theories

The contexts outlined above are important as a backdrop against which to understand changing intellectual ideas about humans and their diversity, in which we can observe the development and consolidation of theories of race towards the point expressed in Knox's statement that 'race is everything'. It is important to grasp that scientific and philosophical thought about race over this period was very diverse: as one might expect, different thinkers had very varied views of how to explain humans and society, and disagreed with each other vehemently. Intellectual and scientific thought had its own internal dynamic of problem solving, which was not determined in simple terms by the external conditions of the world in which it took place.

Of course, internal and external processes cannot be separated: the encounters of Europeans with new worlds posed new theoretical problems; the fact that Europeans were dominating large areas of the world and enslaving millions of Africans stimulated debate about why this was happening; finally, it is impossible to ignore the simple reality that coherent and encompassing theories of race, which divided humanity into natural units and consistently placed Europeans and whites at the top of a global hierarchy, and explained this in terms of natural processes, were developed during a period in which Europeans came to dominate economically and politically almost half the world.

However, we know that imperialism and enslavement had already taken place without this particular nineteenth-century version of racial thinking, so we cannot explain the latter simply in terms of the former. Also, there is no precise fit between specific scientific theories and external contexts: for example, the development of racial typological theory in the 1850s and 1860s (see below) preceded the most active period of European imperialism from the 1870s. The point is that we cannot reduce the complexities of intellectual racial thinking to economic and political contexts (Anderson 2006), even if it is legitimate to see broad correlations.

Underlying the diversity of racial thought over this long period, we can usefully distil out certain broad movements (Banton 1987; Hannaford 1996; Smedley 1993; Stepan 1982; Stocking 1982).

From environment to biology

Explanations of human diversity had long used climate and geography (and the heavens) as external causes, as we saw in Chapter 2. Such environmentalist approaches flourished in the eighteenth century and became more independent of both astrology and theology (Glacken 1967: 620). Climate was not linked only to human physical variety and health, but to political and moral orders. For the French philosopher Montesquieu, climate and environment influenced the 'temper of the mind', the passions of the heart', and the form of political government as much as the physical shape of people. In his work *Spirit of the Laws* (1748) he used the word race infrequently, and then in reference to races of Franks and Visigoths, seen as stages of political history rather than natural varieties of people. Still, the absence of a coherent theory of racial biology did not prevent him from anti-black prejudice, nor from justifying black slavery partly on the grounds of the physical nature of Africans. Reasoning that it is 'natural to look upon colour as the criterion of human nature', he wrote: 'It is hardly to be believed that God … should place a soul … in such a black ugly body … It is impossible for us to suppose these creatures [the blacks] to be men, because, allowing them to be men, a suspicion would follow that we ourselves are not Christians [because we have enslaved them]' (cited in Hannaford 1996: 198–9).

David Hume, in his essay 'Of National Characters' (1742), rejected climatic explanations – what he called physical causes – for differences between 'nations'. Instead, differences were determined by 'moral causes', that is, the social environment, including the 'nature of government … public affairs, the plenty or penury in which people live … and such like circumstances' (Hume 1987: I.XXI.2). These circumstances determined and fixed the character of a people, or even a profession, in a habitual and engrained way. However, when talking of nations which lived 'beyond the polar circles or between the tropics', Hume qualified his own theory, stating that 'there is some reason to think, that [these nations] … are inferior to the rest of the species, and are incapable of all the higher attainments of the human mind'. When it came to 'Negroes', in particular, Hume's doubts increased and he was 'apt to suspect the negroes to be naturally inferior to the whites': even the most barbarous of whites, such as the Tartars, had 'something eminent', whereas among blacks there were 'no ingenious manufactures … no arts, no sciences'. 'Such a uniform and constant difference', in his view, 'could not happen, in so many countries and ages, if nature had not made an original distinction between these breeds of men' (Hume 1987: I.XXI.20, note 10). Like Montesquieu, Hume rarely used the word race, and when he did so he referred to a race of princes or even a race of financiers (1987: I.XII.13); yet, even though, unlike Montesquieu, he rejected environmental explanations, strong elements of racial thinking crept in when he discussed black people.

Towards the end of the 1700s environmentalist explanations began to weaken. Thomas Jefferson, the second president of the United States, for example, was uncertain about the differences between blacks and whites. On the one hand, he believed in the equality of all humans before God; he also adhered to the environmentalist explanations common at the time and used these to explain the condition of Native Americans. On blacks, he wrote, in *Notes on the State of Virginia* (c. 1781): 'The opinion, that they are inferior in the faculties of reason and imagination, must be hazarded with great diffidence' because such a conclusion 'would degrade a whole race of men'. Yet, reflecting on slavery, he also opined of blacks that 'it is not their condition then, but nature, which has produced the distinction'. Nature, for Jefferson, included both body and mind, as 'moral sense' was part of a person's 'physical constitution': typical of eighteenth-century beliefs, there was no simple division between what we would now call nature and culture. Jefferson noted that 'races of black and of red men ... have never yet been viewed by us as subjects of natural history' – although he was writing precisely at a time when humans were increasingly being seen as objects amenable to the science of natural history – so he advanced as a 'suspicion only, that the blacks, whether originally a distinct race, or made distinct by time and circumstances, are inferior to the whites in the endowments both of body and mind' (cited in Jordan 1977: 439, 489).

Jefferson's contemporary, the preacher and philosopher Samuel Stanhope Smith, had no such doubts: all humans were naturally the same and it was only 'the abject servitude of the negro in America' that condemned him to 'perpetual sterility of genius' (Jordan 1977: 443). Environmentalism such as this – which could cause 'perpetual sterility' – was not necessarily much different from positing 'natural' differences (and of course environmentalist explanations are themselves naturalising ones), but it was an important difference in principle: if human characteristics were caused by environment then there was the possibility of change. The entry for 'Negro' in the 1798 edition of the *Encyclopædia Britannica* did not explain the precise causes that underlay what it described as 'this unhappy race', in which vices had 'extinguished the principles of natural law', but some theory of racial degeneration was apparent in the view that 'Negroes' were 'an awful example of the corruption of man when left to himself' (cited in Eze 1997: 94).

The environmentalist explanations typical of the eighteenth century – whether they emphasised climate or social surroundings – allowed naturalising views of human diversity and, when it came to reflecting on black people in particular, this option was pursued with greater intensity, such that black people were often seen as fundamentally different in their very nature. Over the course of the nineteenth century environmentalist views lost ground to approaches that focused more on the innate biology of the human body, which was held to be relatively fixed and permanent (although not as fixed as is often supposed, as we will see later). The French naturalist Georges Cuvier, for example, was an important influence in shifting attention towards the classification of humans in terms of their physical anatomy, with the skull occupying pride of place in the search to delineate races of humans, defined in terms of permanent inherited bodily traits. Cuvier wrote recommendations to travellers on how to record the physical traits of the peoples they encountered, and urged them to obtain cadavers 'in any manner whatever' and then to prepare them carefully, boiling the bones in chemical solutions (Stocking 1982: 30).

Environmentalism is a naturalising form of explanation: the natural surroundings imbue and engrain people, or whole peoples, with physical and moral characteristics, which, in 'soft' theories of heredity, can be passed on to offspring; these are therefore also theories about the body and, in effect, biology. The shift towards a more 'biological' theory must be seen in this context: it is not a move from 'environment' to 'nature', but from one form of naturalising theory, in which the weight is given to external natural causes that shape body and mind, to another, in which the weight is given to internal biological causes that have their own determining dynamic.

From lineage to type

We have seen that race involves ideas about filiation, blood, descent, heredity and genealogy. Race as lineage focuses on the notion that people are connected by virtue of their descent. A sixteenth-century reference to the 'race of Abraham' was understood to comprise all the descendants of Abraham, no matter what they looked like – and they might have been diverse, as he had two wives, one of whom was Ethiopian (Banton 1987: 30). We know from Chapter 2 that reference to descent could already be used to think of collectives of people (Jews in Spain, blacks in the New World), but during the 1700s and especially the 1800s there was an increasing tendency to divide humans up into types, less on the basis of ancestry and more on the basis of their physical traits (although these were always linked to character and behaviour).

Various late eighteenth-century naturalists – such as Linnaeus and the Comte de Buffon – produced classifications that divided humanity usually into five or six varieties, based on criteria of appearance, geography and behaviour. In his essay on 'Variétés dans l'espèce humaine' (1749), Buffon described the 'race' of Laplanders as having 'broad faces and flat noses … hair black and straight, and skin of a tawny colour', but he also described their customs in detail and it was the fact that the different 'tribes or nations' of Laplanders resembled each other 'in form, in shape, colour, in manners, and even in oddity of customs' that made them 'of the same race of men' (Buffon 1807: 191–8). In 1775 the German naturalist Johann Blumenbach divided humanity into five basic varieties (which he later labelled Caucasian, Mongolian, Ethiopian, American and Malay) and used skull measurements to substantiate this. But he saw humans as all belonging to one species and the differences between the types as uncertain and anyway not fixed. By the late nineteenth century medics and physical anthropologists were measuring skulls in very detailed ways in an attempt to fit them into a simple schema of racial types, seen as more fixed, permanent and radically different.

The US medic Samuel Morton followed the lead of Cuvier and Blumenbach in amassing a collection of skulls, and he focused on measuring their internal capacities, showing to his satisfaction in an 1839 publication that Caucasians had the largest skulls, which he took as indicating the greatest capacity for civilisation – even though the sample skulls came from 'the lowest class' of society. Africans were at the bottom of his scale (Banton 1987: 35). Morton ignored the fact that skull size is related to body size, but more importantly he simply assumed that human brain size relates directly to intelligence or 'civilisation'. Followers of Morton included the medic Josiah Nott and the Egyptologist George Gliddon, who, in *Types of Mankind* (1854), argued that, although all contemporary races were more or less mixed, the main physical characteristics of a racial type persisted through long periods of time, such that a type could 'outlive its language, history, religion, customs and recollections'. They also argued humanity was divided into a small number of racial types, which were ancient and permanent. In their view Caucasians were superior: they had the highest intelligence and '*theirs* [was] the mission of extending and perfecting civilisation' (cited in Banton 1987: 41, 43).

The most separate races – such as blacks and whites – could not mix well sexually and were liable to produce weaker hybrid offspring. This was because the racial types were defined as species within the genus *Homo*. (Members of different species, as defined by biologists, cannot produce offspring or – as is the case of horses and donkeys – produce infertile offspring.) Defining races as – or tantamount to – separate species was the most rigid expression of racial typological science. Although Nott had observed very fertile families of mixed-race people in the United States, he explained this away by conjecturing that the mixtures were between races he considered more proximate – the Europeans involved were from southern Europe and darker skinned – which were more able to mix productively.

From monogenesis to polygenesis

The questions of type and species are closely linked to the growing presence of theories of polygenesis – the idea that the types, races or species of humankind had distinct remote origins. The theory went against monogenesis – the idea that all humans had the same remote origin – which had long been the orthodox view, and one supported by Christian theology. Polygenism could be made to square with monogenism to some extent by arguing that, although all humans had the same origin in the very beginning (which, by the mid-1800s, geologists were contending was a lot more remote than the 6,000 or so years deduced by seventeenth-century theologians using biblical timelines), they had soon diverged into distinct groups, which had persisted ever since. Still, the theory of polygenesis was a key component of what Anderson (2006: 110) calls the 'profound intellectual shift' away from Enlightenment ideas of humans as fundamentally the same in nature.

Anderson argues that there are no simple connections between a colonial context and the development of these new innatist racial theories: the context remains fairly similar in many respects, but the theories take on new dimensions. She argues that European encounters with Australian Aborigines played a crucial role in pushing racial theories in the direction of innatism and polygenesis. Aborigines were not just a standard-issue colonised 'other', whose oppression was facilitated by theories of racial difference and inequality; they had specific features that puzzled European observers. In particular, Europeans were concerned by the fact that Aborigines did not appear to settle on, cultivate and improve the land. This was understood to be a defining trait of humanity: human reason led people to do this; it was in human nature to rise above a simple state of nature by improving the surroundings. Yet Aborigines did not do this, as far as the colonists could see, using their own criteria of what counted as improvement, which focused on settlement and cultivation; neither did Aborigines seem to respond to colonial efforts to persuade them in that direction. The apparently peculiar nature of the Aborigines was one element that helped convince theorists that some types of people were not really human, or at least were innately, permanently and radically different from – and inferior to – Europeans.

The impact of Darwin and evolution by natural selection

Darwin's *Origin of Species* (1859) appeared in the heyday of racial typological theory, and its underlying logic contested some of the basic principles of the theory. Darwin argued that all humans had a common origin. He also said that all species were in a continuous process of change, as the environment created pressures that, over time, selected in favour of specific traits that allowed individuals in a population to reproduce more effectively; these traits would become more pronounced and more pervasive in that population. The idea of fixed permanent types did not work.

Yet Darwin himself thought that the races of man had been formed in a very ancient period, and differed then more or less as much as they did now: traces of polygenism thus persisted (Stocking 1982: 46). He weighed the arguments for and against treating races as species and concluded that, although the different races were different in some respects, and although a naturalist seeing an African and a European for the first time might classify them as different species, they were not in fact species. But he retained the basic notion of race as a way of talking about human diversity.

Evolutionist thinking was also able to adapt to the idea of a hierarchy of races (Anderson 2006: ch. 5; Stocking 1982: ch. 3). Darwinian theories were (and are) non-teleological: natural selection does not lead in any predetermined direction; it is not goal oriented. Whatever trait creates a reproductive advantage for existing individuals in a given environment will be selected for over time in the population. Racial thinking – and Western philosophy more generally – tended to be highly teleological: it was assumed that the direction of evolutionary change was towards improvement and civilisation, a goal which put Europeans and whites at the top of the ladder, in their view. Other races or peoples had become stuck on that ladder, whether for environmental or biological reasons. Indeed, ideas that Darwinian approaches were based on 'the survival of the fittest' – a misleading shorthand for the complex process of evolution through natural selection – sustained the concept of a hierarchy in which 'fitter' races could and should dominate over less fit ones. Fitness could in part be judged by power and dominance, resulting in a circular argument: the powerful were fit and the fit were powerful.

Important here, too, was that Darwinian thinking deepened the shift, outlined above, towards understanding human diversity – physical and social – in terms of biological processes. Especially in the guise of the social evolutionism of Herbert Spencer – who coined the phrase 'survival of the fittest' and applied ideas of evolution to whole societies, not just biological populations – the notion that the laws of nature underlay all aspects of human development became more powerful. For all these reasons, then, the basics of racial science persisted into the late nineteenth century and beyond.

Sara Baartman: a case study

A telling and controversial example of the shifts in scientific approaches to race in the nineteenth century is the case of the so-called Hottentot Venus. The case caused some debate at the time and has been the subject of numerous commentaries, which use it as a window on contemporary racism, racial science and the sexual objectification of black women (Crais and Scully 2009; Gilman 1985; Magubane 2001; Qureshi 2004).

The story is that, in 1810, a Khoisan woman from South Africa – a Hottentot or Bushman, in the imperial language of the time – was brought to England by a doctor who doubled as an exporter of museum specimens. She was known as Sara Baartman, a name probably given to her by her Dutch master in the Cape region. She was sold to an entrepreneur who put her on public display as an exotic specimen in the context of London's entertainment scene. Although London was by then home to a small black population, mostly of American origins, as an African she attracted some attention. Viewers were interested in a feature said to be typical of Khoisan women, namely steatopygia, a large fat accumulation on and around the buttocks; this feature was visible to the public as Baartman was dressed in very tight clothing; indeed, this caused scandal among those who thought it indecent. Khoisan women were also reputed, at least among medics and natural historians, to have elongated genital labia, the so-called Hottentot apron; members of the public could not see this for Baartman, however. In England abolitionists took up her cause – slavery had been abolished in England in 1807, but continued in the empire until 1833 – and attempted, unsuccessfully, to get her repatriated to South Africa.

Baartman was exhibited in other locations in England – and apparently married a groom at some point – and was then exhibited in 1814 by an animal trainer and showman in Paris, where her show was very popular and ran for eighteen months. During that time she was examined by the naturalist George Cuvier and his colleague Geoffroy Saint-Hilaire; she posed nude for them, and detailed drawings of her were made and appeared in the text *Histoire naturelle des mammifères* (1824–37) (see Fig. 3.1). Baartman died of an infection soon after, and Cuvier took casts of her body, made a wax mould of her genitalia, dissected her body and preserved her skeleton; he produced a detailed report of his findings. The cast and skeleton were displayed in the Muséum national d'histoire naturelle and then the Musée de l'homme in France until the 1970s. In 1995 a campaign started to have the remains repatriated to South Africa, which occurred in 2002.

Shocking as this is to today's sensibilities, the public exhibition of people seen as odd or unusual, and among them, of 'exotic specimens' (indigenous Americans, for example), had a long history in Europe; it became more common in the imperial late nineteenth century with human zoos, world fairs, international exhibitions and museums displaying colonial peoples. The scientific interest in and dissection of bodies was also common and growing in scale, as we have seen. Various travellers and natural historians had written about Khoisan steatopygia and genitalia before Cuvier. In this sense Baartman was not unusual, although her case has achieved iconic status. But it serves as a good example of a number of key features of nineteenth-century racial thinking.

First, we can see the intense focus on the body and its anatomy as the location of clues to racial difference, to be discovered through meticulous scientific observation. Cuvier saw humans as all one species, but on the basis of her anatomy and physiognomy he classified Baartman as of the 'Boschiman race' (from the Dutch *bosjesman*, Bushman), a sub-branch of the Ethiopian or black race in contemporary classifications, and the lowest level of humanity in his view, close to the apes. Second, although the Khoisan were

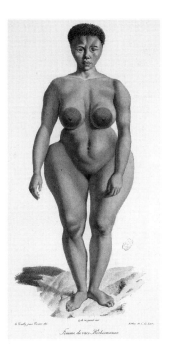

Figure 3.1. Sara Baartman, from *Histoire naturelle des mammifères*. The use of this kind of image is controversial: see the 'Further research' section.
(© RMN–Grand Palais/Muséum national d'histoire naturelle)

seen as a specific race and not typical of all black people, the specific role of blackness as a counterpart to whiteness is made clear: as for Montesquieu, Hume and Jefferson, blackness figured in racial theory as a kind of baseline reference point, the place where doubts crept in as to the unity of humankind. Interestingly, at the time, English scientists often linked the Irish with Africans, despite the difference in skin colour: both were seen as inferior, and this was explained by seeing the Irish as biologically linked to Africans (Magubane 2001).

Third, some of the concern with race, anatomy and blackness was also a concern with sex, in keeping with a Foucauldian analytics of sexuality. The proper management of life was threatened by forces of sexuality seen as excessive or anomalous. Colonial populations, the lower classes and women in general – especially prostitutes – were seen as the locus of such anomalies; black men and especially black women were stereotyped as over-sexed, and seen as dangerous, but also fascinating. The focus on Baartman's buttocks and genitalia was an expression of this preoccupation with a powerful interweaving of race and sex, in which the aim of regulating life force and keeping sexuality – especially female sexuality – within 'proper' bounds coalesced with the aim of keeping white nations and races free from racial contamination and maintaining them – especially white men – in a superior position. Lastly, the entire episode conveys very powerfully the sense of colonial and class superiority of the scientists and exhibitors, and the entitlement they felt to use the bodies of colonised people for their own purposes – to observe, objectify and dissect.

3.3 The spread of racial theory: nation, class, gender and religion

The theories I have been outlining had the weight of science behind them. If Jefferson complained in 1781 that people – or at least the 'races of black and red men' – had not been made the subjects of natural history, then during the 1800s natural history, or rather the new biological science, was taking a very profound interest in these and other races. We now know these racial theories to be unfounded – a blatant expression of white supremacy – but at the time they had the status of scientific knowledge. There were debates about the theories, as there usually are among scientists, but this did not diminish the basic authority of science itself.

These racial theories were elaborated mainly in Europe and the United States, and they tended to focus on the four or five main races identified by theorists and scientists – although each race could be subdivided into numerous smaller units, as in William Z. Ripley's *Races of Europe* (1899) or Cuvier's Boschiman race. But the influence of these theories was global in reach: we saw in Chapter 1 that the British used racial theory to interpret society in India (Bayly 1999), while Chinese thinkers also used Western ideas about race, albeit selectively and via only a few translations (Dikötter 1992); below, I give an example from Japan. Race theory could also be applied to the understanding of other social groupings, such as the lower classes and Jews, as I show below.

Japan: nation, empire and race

A good example of the influence of European racial theory on other areas of the world comes from Japan (Weiner 1994, 1995). As noted for China in Chapter 1, there was no shortage of long-standing ideas about the superiority of Japanese culture and people, but these were given a distinctive inflection by European ideas. Concepts of social evolution, as propounded by Darwin but especially by Spencer, were introduced to Japan by Edward S. Morse, a US zoologist and student of Louis Agassiz, a well-known proponent of polygenism and of the idea that races were species. Morse went to Japan in 1877 to collect shells for his research and was offered a job as a professor of zoology at the Tokyo Imperial University. His lectures on evolution and translations of the works of Darwin, Spencer and Thomas Huxley all had an important impact. The idea that different races – or, more generally, groups defined by blood, descent, physical type and culture – had different innate capacities, which allowed some to be superior to others, worked in late nineteenth- and early twentieth-century Japan in a number of ways.

First, the idea of race fitted easily with emerging patterns of Japanese nationalism, linked to processes of modernisation and nation building taking place under the Meiji regime (1868–1912). The nation was portrayed as a family, linked by blood and ancestry, and a natural community in which cultural, historical and physical aspects all coalesced. *Minzoku* was a term popularised in the 1880s to refer to the nation: often glossed now as 'ethnic group' and apparently referring to culture alone, the term – like the English 'ethnic' – in fact had strong connotations of shared blood, ancestry and physical substance. This concept of the Japanese as a race-nation was reinforced by victories over China (1894–5) and Russia (1904–5); it also legitimised Japanese colonial control of Korea and Taiwan, which started in 1895. Although these people were, from a European point of view, all part of the same race, the notion of a specifically Japanese *minzoku* gave credence to the idea that the Japanese were superior and destined to rule over others in the region. At the same time, imperialism was seen as a way to sustain the race: according to contemporary historian Tokutomi Soho it was a 'policy born out of necessity if we are to exist as a nation and survive as a race' (cited in Weiner 1995: 451). It is notable that the concept of a Japanese *minzoku* relied on ideas of kinship and blood: physical appearance did not play a very important role. Lightness of skin was a valued aesthetic attribute, especially in women, and there was some idea that non-Japanese Asians and some 'outsider' minorities in Japan, such as the Burakumin, were darker in colour. Meanwhile, Chinese war enemies were portrayed as pigs and 1930s Chinese caricatures portrayed the Japanese as particularly hairy (Dikötter 1992: 141). But the key role played by skin colour in European ideas about race did not operate in the same way.

Second, the possibility of a racial kinship between different Asian nationalities also permitted the idea of an Asian or yellow race – and here skin colour did play an important role – distinct from a white or European race, which were seen as being in competition on a more global scale. The assimilation of Korean and Taiwan people to Japanese ways, which was a policy forced upon the colonies, was also thought to be easier due to racial affinities between these Asian peoples.

Third, the image of a coherent Japanese race-nation was used to legitimise social hierarchies within Japan. Various groups – the urban poor, the dispossessed peasantry, outcast groups – were seen as culturally and innately different from the mainstream dominant classes. A doctor's assistant's account of life in a 1920s rural community described peasants as being 'as black as their dirt walls', as having a 'peculiar odour' and a 'peculiar physiognomy': the peasant could be 'identified by his ignoble face' which was 'common and vulgar' (Hane 2003: 35). The Burakumin outcast group was seen as impure and of alien origin, and subject to extensive legal exclusions until 1871 (Takezawa 2005).

Race, class and gender

The use of racial idioms to talk about class differences was not peculiar to Japan. A well-known example is the way the Irish were described in nineteenth-century England. They had long been portrayed by the English as barbarians and savages (Smedley 1993), so it is no surprise that nineteenth-century racial theory would provide the means to describe them as a race, conceived now as a natural community (Curtis 1997; Miles 1989: 58); as noted above, some racial theorists also linked the Irish to Africans.

Anne McClintock describes similar processes in late Victorian Britain, but in ways that bring in an important gender dimension. She argues that British society of the period was characterised by a 'cult of domesticity', which confined women to the home and gave men pride of place in the public sphere of political power and the market. At the same time, such a cult was only conceivable for middle- and upper-class households – and even then only in some ideal patriarchal order – because poorer women routinely worked outside the home; furthermore, the homes of the middle and upper classes depended on such women's domestic labour as servants. McClintock argues that one response to this contradiction between the ideal image of the woman and the reality of her paid labour – happening right under the noses of those propagating the idealised image of the woman – was to racialise these working women as primitives; to project on to them the image of the dark-skinned colonised other and thus make them seem more distant from the moral order of British society at home.

McClintock analyses the drawings of a Victorian barrister, Arthur Munby, who depicted working women in ways that clearly assimilated them to stereotypical images of blackness (see Fig. 3.2) and she analyses too how Munby's long-term relationship with his maid, Hannah Cullwick, which he documented and photographed in detail, was replete with racial images, including depicting her as a slave and blackened with dirt (McClintock 1995: 75–131). Munby admittedly was a sexual obsessive but, in the way he tried to manage the contradictions of class and gender of Victorian society, he 'was no eccentric, but

Figure 3.2. Drawing by Arthur Munby of himself and a female miner
(Master and Fellows of Trinity College Cambridge)

was fully representative of his class' (McClintock 1995: 80). The fact that black people were seen, in late Victorian racial theory, as a separate species and a natural kind separate from whites gave this kind of racialised othering a peculiar force.

Anti-Semitism and biological racial theory

The idea of race was applied to Jews in the late nineteenth century, and proved powerful and, ultimately, very destructive. As we saw in Chapter 1, anti-Jewish prejudice had existed since medieval times in Europe, but through the 1600s and 1700s there was no theory of a Jewish race as a biological entity, despite Spanish references to *raza de judío* (Jewish blood or ancestry), as contaminating a person's lineage. Such biological theories of Jewishness really emerged after about 1870, but they were prefigured, especially in Germany, by ideas about the cultural and spiritual particularity of European peoples, enshrined in the notion of a *Volk* (people). The eighteenth-century German philosopher Johann von Herder had developed the concept of a people sharing a *Volksgeist* (folk spirit or soul), united in a spiritual affinity for their common language, history, religion, education and tradition – in a word, their *Kultur* (culture). Although he firmly believed that humans were all one species – in accordance with much thinking at that time – he nevertheless saw culture as deeply imbued and more or less innate in the people who inhabited the physical space of the nation and were heirs to the *Volksgeist*; each people had its own spirit, and other cultures were a source of contamination.

Such thinking gave a powerful rationale for discriminating against those seen as not true offspring of the nation and its soil, and Herder saw Jews as a people out of place in Europe. Assimilation by Jews was considered necessary, but there were always doubts about whether it was really possible: Jews seemed to retain their own identity as a specific ethnic and religious group, even as they became increasingly bourgeois. During the late eighteenth and nineteenth centuries laws restricting Jews and confining them to urban ghettos generally relaxed in France, Britain and Germany, but powerful suspicions remained and grew stronger in reaction to this partial emancipation. Jews were seen as undermining the German *Volksgeist*, siding with the rationalist and secular spirit of the French Revolution against German Romantic spirituality and Christian values. Jewish financiers were suspected of having been instrumental in the defeat of France in the Franco-Prussian war of 1870, aiming to plunder the country. On the unification of Germany in 1871 Jews were made full citizens, but anti-Jewish pamphlets and tracts were already being published, one of which, by William Marr, coined the term anti-Semitism. In 1879 Marr established the Anti-Semitic League, which was influential in many European countries.

From about 1880 full-blown biologically racist anti-Semitism took shape, especially in Germany. It was informed by current racial science and argued that the Jews were a biologically inferior race who were plotting to take over the country or undermine it. The German philosopher Eugen Dühring, for example, described Jews as a different species, which had to be extirpated from society. In German publications Jews were not seen as intellectually inferior; instead, they were characterised as evil, cunning, pushy, hyper-rational, too modern; they were depicted as unable to become properly German and fit into the traditional, spiritual, romantic, honourable *Volksgeist* of the nation (Frederickson 2002: 75–80).

Racial theory, then, was not confined to thinking about colonial difference, but informed ideas about religion, ethnicity and class differences. All of these could be phrased in the biological idiom offered by racial theory and by biopolitical preoccupations with regulating the life force of the population in a 'proper' manner, to ensure a nation of sound moral and physical fibre.

3.4 Nature, culture and race

The emphasis on biology and anatomy that characterises nineteenth-century racial science is an important shift, but in my view some care is needed with how this is interpreted. Of course, racial science was never just about biology. The whole point was not simply to divide humans up into races or species on the basis of their biology, but to use this to explain psychological and behavioural traits, and the perceived differences in the level of 'civilisation' achieved by various peoples. As usual, race was about nature and culture. But in looking at nineteenth-century ideas about biology and nature, it is too easy to assume that we know exactly what was involved here – that 'nature' and biology were all about fixed, permanent and immutable characteristics. Things were not so straightforward.

To start with, the racial type that was supposedly defined by fixed biological traits was explicitly recognised as an abstract, underlying reality. Thus Nott and Gliddon acknowledged that all contemporary peoples were shaped by 'climate, mixture of races, invasion of foreigners, progress of civilisation, or other known influences', but that the underlying type could somehow be inferred from the superficial appearance (cited in Banton 1987: 41). Likewise the sociologist William Z. Ripley, in his *Races of Europe* (1899), posited three basic racial types in Europe, which he deduced from maps and graphs showing the frequency and distribution of certain traits in the region: 'Finding these traits floating about loose, so to speak, in the same population, we proceed to reconstitute types from them.' Any individual might not show all the features of a given type, but this could be explained by the influence of 'chance, variation, migration, intermixture and changing environments'. In fact, the racial type was not only abstract but 'unattainable' (cited in Stepan 1982: 94; Stocking 1982: 62–3). True races had been formed long ago and then become altered and mixed over time. Even the skull, which Ripley saw as least likely to change, was subject to environmental modification. In that sense, these racial theorists acknowledged the impact of the environment; the biology that remained fixed, immutable and unchangeable was actually a theoretical ideal.

The role of the environment was also acknowledged in the widespread and persistent belief in the inheritance of acquired characteristics and more generally theories of 'soft' inheritance, which held that the heredity material passed on from parent to offspring could be influenced by the experiences and milieu of the parent. By the 1870s and 1880s scientists such as Francis Galton (the founder of eugenic theories – on which more below) and August Weismann were putting forward theories of hard inheritance, in which the 'stirp' or 'germ plasm' passed unchanged from one generation to the next; these theories would be confirmed by twentieth-century genetics. The theories also gave a particular shape to a conceptual difference between nature and culture, understood as an opposition between a natural, biological, given substrate and a cultural overlay of learned behaviour.

Until then and, indeed, through the 1920s in some medical circles, ideas of soft inheritance continued to command adherence and to blur a clear distinction between nature and culture. In particular, many medics posited mechanisms of degeneration – illness, alcoholism, sexual excess, even urbanisation – which would weaken and undermine the bodily substance of heredity; the 'stigmata of degeneration' would be revealed in the physiognomy of the individual (Lubinsky 1993). This medicalised notion of degeneration built on earlier ideas, going back to Buffon, of a decline away from a more perfect time and form, and it extended ideas about hereditary problems in families to encompass the condition of entire nations and races in Europe and other countries, such as Brazil (Borges 1993).

Racial contagion in South East Asia

An excellent example of the soft line between nature and culture and of the need to grasp the 'biology' of nineteenth-century racial theory in its historical context comes from Stoler's studies of Dutch colonies in South East Asia (Stoler 1995, 2002). One of Stoler's starting points is that the categories of coloniser and colonised were by no means clearcut. Although racial theories about fixed and immutable biological differences might seem to provide an easy way to make the distinction, in practice both categories had to be constantly recreated, regulated and maintained through strategies of colonial rule, which used ideas about race, colour, internal essence, education, sexual behaviour and religion to enact the categorical differences and define who was 'white' and who was 'native'. Meanwhile, the boundaries were continually contested by sexual relations, the birth of mixed-race children and the intimacies of colonial life in which native servants played a key role in domestic contexts. In the 1880s in the Dutch East Indies nearly half the European male population was unmarried and living in concubinage with Asian women, a practice that at the time was actively condoned by colonial authorities as useful and necessary for these men; children from such unions were considered European, but there were fears about them, as they not only had 'native blood' but were brought up by native mothers and/or servants. As more white women took up residence in the colonies in the early 1900s, open concubinage declined and greater segregation was considered necessary to protect white women's morals.

However, threats came not only from colonised men, women and mixed-race children encroaching on white European domains and rights, but from the possible degeneration of the whites themselves, especially the poorer class of whites. The tropics had the power to undermine European colonists. In 1856 a Dutch civil servant said the assumption that a 'European is a European and will remain so wherever he finds himself' was not correct:

the tropics meant that a European in the Indies was 'an entirely different being than in his country … he no longer can be considered as a European, at least for the duration of his stay in the Indies, but rather as belonging to a specific caste of the Indische population'. In 1898 a Dutch lawyer concluded that, for whites who remained for years in native surroundings, the milieu had 'the power to almost entirely neutralise the effects of descent and blood'. Later in 1907 a Dutch doctor wrote that a European born and raised in the Indies, in the absence of a properly controlled environment, could 'metamorphose into Javanese'; the physical constitution of children brought up by native women, who spoke to them in a native tongue and breast-fed them, could be at risk. As one writer put it in a 1941 text on life in the Indies: 'A Dutch child should grow up in Holland. There they will acquire the characteristics of their race, not only from their mother's milk but also from the influence of … the fatherland.' Although perhaps only metaphorical, the reference to mother's milk harked back to much older beliefs about breast milk as having the power to shape a child's constitution and blood. In 1932 one French academic said of the white colonial: 'The climate affects him, his surroundings affect him, and after a certain time, he has become, both physically and morally, a completely different man' (cited in Stoler 1995: 104–5; Stoler 2002: 66, 74–5, 98). Physically, such people were prone to a range of illnesses, including fatigue, physical breakdown and racial degeneration. Morally, these colonials would, in the eyes of the authorities, tend to neglect the conventions of superiority and become ostentatious, inactive and demoralised, faults they absorbed from the local population.

All this gives a different gloss on the idea of biological determinism and the 'immutable' racial traits of nineteenth- and early twentieth-century racial science. There was a flexible idea of where the boundary between biology and culture lay; great weight was attributed to the influence of social and physical surroundings on a person's moral and physical constitution; and there was concern about the durability and immutability of a person's nature.

Eugenics

Eugenics is often seen as one of the clearest expressions of racial science, based on the idea of controlling the biological breeding of the nation's population to achieve a stronger and fitter race; such control could include the forcible sterilisation of people deemed 'feeble-minded' or otherwise 'unfit'. Under the Nazis such eugenic ideas developed into doctrines of Aryan supremacy and campaigns of extermination carried out against Jews, 'Gypsies' and others seen as racially inferior.

The idea of eugenics was created by Francis Galton, one of the first to propose a hard theory of inheritance. Yet Galton defined eugenics as 'the study of the agencies under social control that may improve or impair the racial qualities of future generations either physically or mentally' (cited in Stepan 1982: 111). Any agency in which people might intervene was considered relevant. Paradoxically, given Galton's hard theory of inheritance, which would suggest that only direct interventions into the transmission of the 'stirp' would have any effect on racial qualities, eugenics had a strong affinity with the notion that acquired characteristics were heritable: social interventions could also improve racial qualities.

When eugenics caught the popular imagination in the early years of the 1900s it was a social movement led by middle-class reformers, who saw themselves as progressive. A whole range of measures were involved, including attempts – usually directed at the working classes – to improve hygiene, domestic environments and mothering skills. It was only a little later that some eugenicists called for laws allowing sterilisation of people deemed unfit, and it was mainly in the United States that such practices took place: some 9,000 people were sterilised there under these laws between 1907 and 1928 (Kevles 1995). In Latin American countries such policies gained no traction, and eugenics was essentially a movement of 'social hygiene', alongside attempts to increase European immigration in order to 'improve the race' (Stepan 1991; Stern 2009). In the East Asian colonies eugenics provided a medical, biological idiom for the kind of beliefs and practices described above, directed primarily at protecting the moral and physical fibre of whites and drawing clearer boundaries around whiteness (Stoler 2002: 61–5).

3.5 Black reaction

How did the people categorised as biologically inferior react to racial theory? The black US intellectual W. E. B. Du Bois started his university studies in 1885, in Nashville: for him the social context was of strict racial segregation and the intellectual context of racial typological science. He moved to Harvard and did a Ph.D. in history and philosophy, with a thesis on the suppression of the slave trade in the United States, having also studied in Berlin for a while. At Harvard he recounted that he 'began to face scientific race dogma'. He did not contest race theory by adducing details of biology and anatomy, but rather noted that the measures used to distinguish races were inconsistent: science had not managed to sort out the 'relative authority of these various and contradictory criteria'. He admitted that people had been separated into groups by 'subtle forces' and while these 'generally followed the natal cleavages of common blood, descent and physical peculiarities', they also at times 'swept across and ignored these'. Races still existed, 'clearly defined to the eye of the Historian and the Sociologist', but the main differences were spiritual and psychical ones, based on physical differences, but 'infinitely transcending' them. Thus 'the Teuton nations are, then, first their race identity and common blood; and secondly, and more important, a common history, common laws and religion, [and] similar habits of thought' (cited in Baker 1998: 110–13).

This slightly tentative critique of racial theory cedes it a certain amount of ground, but nevertheless subordinates the physical criteria, which were primary for the race theorists, to cultural ones – a substantial departure from orthodox views in the late 1890s. Du Bois's detailed ethnographic and sociological study *The Philadelphia Negro* (1899) was perhaps more incisive, because it argued that all the problems that white residents identified as being located in the black neighbourhoods, and which, in their view, affected the governance of the city as a whole, were due to poverty, segregation and poor health care, rather than to racial inferiority, degeneration and innate dispositions towards criminality (Baker 1998: 114). In short, then, the main response was to separate the biological from the social and to focus on the latter, a trend that was to gradually strengthen during the twentieth century, despite the persistence of eugenic thinking into the 1940s.

A similar response can be seen in Latin America, where typological racial theory and hardline innatist eugenics were less powerful anyway. In Cuba, for example, the Independent Party of Colour was formed by Afro-Cubans in 1908 in response to what they felt was their exclusion from politics and public office, despite their crucial role in the wars that had led to independence from Spain in 1898. Their agenda was not primarily about biological difference, despite the fact that at this time academic and popular views of Afro-Cubans were informed by the racial science of the day and tended to see them as inferior and infantile. The Cuban sociologist Fernando Ortiz, for example, produced a major study of Afro-Cuban criminal behaviour, much of which was, he said, spurred by Afro-Cuban *brujos* (sorcerers, 'fetishists'). A dedication to the first volume (*Los negros brujos*, 1906) was contributed by the Italian medic Lombroso, well known for his criminological theories, which proposed that criminals could be distinguished by heritable physical characteristics, particularly on the skull and face. Ortiz claimed that 'fetishism is in the mass of the blood of the black Africans' (Helg 1995: 112).

Afro-Cuban leaders contested racism, based on ideas of black inferiority, as 'artificial' and 'barbarian' (Helg 1995: 43). There was also some questioning of scientific racism by asserting that the unity of the human species was not only a theological but also a scientific truth; skull shape was not a consistent measure for defining race and could not prove the inferiority of some races (1995: 149). However, the main platform was social equality – education, land, jobs. Like Du Bois, then, black critiques in Cuba did question biological theories but preferred to shift the debate onto the terrain of social relations and the rights due to citizens in a liberal political order.

CONCLUSION

This chapter has traced a series of gradual transitions in which science began to understand humans as biological beings, slavery was abolished and replaced by free and semi-free labour in an increasingly capitalist and industrial global economy, European empires were vastly extended, nationalism flowered, liberalism as a political order became more widespread and had to square the circle of equality and hierarchy, and governance became concerned with the life forces of the population. During this long nineteenth century racial thinking also moved towards biology and anatomy as the location for understanding human difference, especially hierarchical difference, and towards a theory of racial types and ideas of polygenesis. Darwin's theories of evolution, while in principle falsifying key components of racial theory of the time, actually circulated as social Darwinism, which for a while reinforced current versions of racial thinking by providing a biological idiom for understanding social as well physical processes of change.

As ever, because racial thinking focused on blood and ancestry – now in biological mode – it continued to intersect with sex, the mechanism for the transmission of blood and racial traits. As sex was also the biopolitical interface between individual and population, in Foucault's terms, it acted as a nerve centre for preoccupations with race, purity and otherness, as we saw in the case of Sara Baartman.

The racial theory of the period had a pervasive impact, meshing with ideas about class and gender difference, and providing idioms for hierarchical difference and imperial conquest in non-European societies such as Japan, where it built on pre-existing naturalising theories of difference. But the immutability of race, which racial theory itself emphasised, although recognising this as an 'unattainable' abstraction, and which many scholars also see as characteristic of race at this time – and, because this period is often seen as the defining epoch of race, as a central feature of racial thinking in general – turns out not to be so straightforward. In fact, we find important elements of the thinking we saw in the previous chapter, in which mechanisms of inheritance are 'soft', 'culture' shapes 'nature' and there is not a clear dividing line between them. Racial thinking does not operate only by fixing things – consigning certain peoples to an immutably inferior position – although this is certainly one of its modes of operation. It can also operate by positing the danger of change – whites may degenerate in tropical climates and in the hands of native servants; or, as we saw in Chapter 2, indigenous Americans might become increasingly Spanish by eating Spanish food. If colonised peoples were really immutably inferior and white colonists completely fixed in their superiority, there would be little cause for concern. But the threat of mutability requires the constant enactment of stringent measures to control and limit this danger and maintain the difference, not only in behaviour but in body.

This is relevant to the relation between racial thinking and colonial practices, including enslavement. We have seen in this chapter that it is difficult to trace specific changes in racial thinking to specific developments in colonial practices. But a broad correlation between racism and colonialism–imperialism seems inescapable. Montesquieu, Hume, Jefferson, Nott, Gliddon and Knox all saw black people as inferior, whatever the particularities of their thinking about human diversity. Jefferson was unsettled by the idea of a natural difference between blacks and whites, because this would 'degrade a whole race of men', but black people in the United States had already been degraded, in social terms, through slavery. It was, after all, in 1781, the date of Jefferson's *Notes*, that the British captain of the slave ship *Zong* threw some 142 living African slaves into the Atlantic ocean in order to prevent the spread of infection on the ship, banking on the fact that the insurers would cover the loss of what were, in commercial terms, 'perishable goods'. Such inhumane acts did not need the intellectual justification of seeing blacks as an inferior species or biological type: existing ideas about the inferiority of Africans did just as well.

FURTHER RESEARCH

1. Use Eze's book (1997) to explore eighteenth-century writers' views on race. What differences and similarities can you see with today's understandings of human diversity?

2. Explore the Sara Baartman case, using the references cited and online. In addition to the image shown, there are many other visual images of Baartman available online, and some would argue that reproducing them can only reinforce the sexual objectification of black women (see also Collins 2000). Do you agree? Artist Ayana V. Jackson's 2013 exhibition, 'Archival Impulse and Poverty Pornography', might be a useful resource here. Some bloggers have recently drawn parallels between rap artist Nikki Minaj and Sara Baartman. Does this make sense?

4 Biology, culture and genomics

In the last chapter we already entered the twentieth century, via eugenics and the continued influence of scientific racial typologies and social Darwinism on thinking about human diversity. This and the next chapter concern changes that began in the early decades of the century, took fuller shape from mid-century and are still with us today. In terms of the conventional chronology of race that guides the first half of this book, and which I am also trying to challenge and refine, this period witnesses a shift from 'biology' to 'culture'. Briefly put, the powerful reliance on biology that characterised late nineteenth- and early twentieth-century thinking about race is generally said to be displaced, especially after WWII, by an emphasis on culture, often signalled by using the terms 'ethnic group' or 'ethnicity' (derived in the 1930s and 1940s from the older term 'ethnic', as alternatives to 'race'). Some approaches describe the complete demise of race – specifically in Europe – as a discredited concept that has lost virtually all traction and been replaced by a 'cultural fundamentalism', in which groups are seen as fundamentally different by virtue of cultures that are deeply rooted and basically conflictive. This fundamentalism does not link cultures to racial biology, nor does it see some cultures as inferior and others as superior (Stolcke 1995). My view is that this does not do justice to the ways race, racial thinking and racial hierarchy do persist, even in Europe.

Others do not envisage the erasure of race, but instead a complex shift, involving different but overlapping spheres – science, politics, everyday cultural practice – and hiding the persistence of race and of many racialised processes of naturalisation, not to mention the persistence of racism, now apparently without the trappings of race or biology. This scenario, in which race is 'buried alive' (Duster 2003; Goldberg 2008), has been said to involve 'neo-racism' (Balibar 1991a), 'new racism' (Winant 2002), 'cultural racism' (Hale 2006: 144; Taguieff 1990) and even 'raceless racism' (Goldberg 2008). Scholars have also referred to 'colour-blind racism' (Bonilla-Silva 2003) and 'race-evasive discourse' (Frankenberg 1993), to indicate the ways many people in the United States, especially white people, try to avoid explicit reference to race, amid claims that we live in 'post-racial' times (Lentin 2014).

In these approaches it is often recognised that culturalist discourses can involve naturalisation, which casts culture as something almost essential and innate, but my view is that this recognition glosses over the complex ways in which nature and culture work together – as they always do – in racial thinking. It also does not do justice to the way ideas about bodies, biology and heredity continue to figure in the racial thinking, and the racism, of this period.

In this chapter I outline some of the key moments and people in the 'retreat of scientific racism' (Barkan 1992), during which the old science of race lost most of its traction: this takes us from Darwin to the Human Genome Project, via both anthropology and biology. I then explore the way race reappears, even as it disappears, in biological science, genomics and medicine, with a brief look too at some psychology and cognitive anthropology. Related to this – as science and society are closely intertwined spheres – in the next chapter I then look at the continued explicit presence of race in public institutional practice, despite the overall trend towards its erasure; and I explore some ways in which biological–natural aspects of race appear in everyday lay discourse, qualifying the idea that we live in an era of simply 'cultural racism'.

4.1 Darwin (again), genetics and the concept of population

In the previous chapter we saw that Darwin's theories went against the grain of racial typology theory, but that aspects of them were assimilated to existing racial theory. However, it is important to appreciate how Darwin's work laid the basis for twentieth-century understandings of human diversity (Banton 1987: ch. 3). Fundamental here was a shift from essentialist to 'population thinking' in biology (Mayr 1982: 45, 119). Eighteenth-century environmentalism saw people's constitutions as shaped by climate and geography and then transmitted by heredity mechanisms; in the process, varieties or races of people emerged. Nineteenth-century typologies thought in terms of permanent types or races that had ancient origins and persisted unchanged over time, albeit as an underlying reality. Both approaches saw varieties or types as collective units, characterised by essential features shared by all members of the unit.

The principles of Darwin's approach mean thinking in terms of diverse individuals, who compete with each other for resources and who are more or less successful as sexual reproducers. The hereditary traits that make an individual more successful, in a given local environment, are passed on to offspring, alongside other hereditary traits s/he carries. Traits for reproductive success are ones that enhance survival and those that make an individual attractive to potential mates. Over time these traits become more common within the population of individuals, and can also become more pronounced phenotypically. In a local environmental niche, then, populations of individuals breeding with each other come to share certain inherited traits, not as an essence, but as a statistical probability. As the environment changes so the distribution of traits in the local population also changes. Environmental niches and their populations are not usually closed systems, but overlap and interchange, which means it can be difficult to define a population.

Darwinian principles lead to a strong connection between geography and heredity, but it is a flexible connection that can change over time and it is a probabilistic one: not all members of a local population share all the same traits; members of different local populations might well share some traits (indeed, we now know that all humans share a vast amount of genetic traits); populations are not easily bounded.

Genetics and population

This notion of population was present in the principles of Darwin's theories, but it was not until the 1930s that it became fully developed, with advances in the newly established discipline of genetics (a term coined in 1905). In the mid-nineteenth century the Austrian monk Gregor Mendel had published findings indicating that heredity occurred by the transmission of 'particulate' traits. Darwin, like others at the time and before, had thought in terms of the blending of two liquid-like flows of heredity material (germ-plasm, 'blood', etc.), so that the offspring would have a kind of average point between the two parents. Mendel showed that specific traits were passed on without blending, that different traits could be inherited independently of each other, rather than as a coherent package, and that they might recombine in different assortments in various offspring of the same parents. That is, particular physical traits were linked to particular genetic 'particles' (or genes as they became known) that maintained their integrity across the generations, even if they were not always expressed in every individual.

This work was ignored at the time, and only rediscovered in the very late 1800s. It made it much easier to see how particular traits were selected for under environmental pressures and became a more frequent feature of the population's biological heritage. Mendel's work also relied on observing statistical regularities in the expression of certain traits over the generations. Darwinian and Mendelian principles, combining probabilistic concepts of population and a particulate theory of inheritance, were a major force undermining the concept of 'racial type'.

4.2 Boas and the separation of biology and culture

A second key force in the dismantling of scientific racism was Franz Boas, a German Jewish scientist, originally trained in physics and then geography, who migrated to the United States in 1887, where he worked as an anthropologist, becoming very influential in the development of the discipline. His work shaped ideas about race in two main ways. First, he carried out anthropometric work on immigrants, showing that head shape – that key component of racial typology theory – could change quite quickly over time as immigrants' children adapted to new diets and surroundings, but without any change in descent (Boas 1912). This undermined the concept of a permanent racial type, using the very tools that had been deployed to sustain it.

Second, Boas strove to separate biology from culture. As we have seen, many understandings of humans prior to this time did not separate the two domains very clearly: when it came to human nature in particular, the distinction between nature and culture became flexible. Things that humans decided to do could influence their very nature; and things that we might now class as culture, such as a person's moral qualities, could be seen as part of their physical constitution. Stocking notes that for social sciences in the early twentieth century the problem was not their domination by notions of 'biological or racial *determinism*, but rather … by a vague sociobiological *indeterminism*, a "blind and bland shuttling" between race and civilisation' (Stocking 1982: 265, citing A. L. Kroeber; emphasis in the original).

Boas wanted to separate out the biological processes of evolution, which underlay humanity's physical diversity, from the processes of social and cultural change, which caused the much greater cultural diversity observed among humans (Boas 1966 [1940]). Boas was not alone here. As noted in the previous chapter, changes in biological thinking from the late 1800s had begun to suggest 'hard' mechanisms of inheritance, in which the hereditary material did not change from one generation to the next; it was not shaped by environment (in the short term) or behaviour. This gave a basis for thinking in terms of a biological substrate, separate from people's behaviour. Other anthropologists also wanted to contest the premises of social evolutionism or social Darwinism, which sought to explain the overall processes of social change in terms of biological mechanisms. The US anthropologist A. L. Kroeber, for example, distinguished between organic and 'super-organic' evolution, with the latter consisting of social and cultural processes which, as it were, lay on top of the organic, biological ones. Like other anthropologists of the time Boas was strongly against the social evolutionary idea that human society had progressed up a ladder of stages, with whites reaching the pinnacle of Western civilisation while others were stuck on lower rungs, a theory strongly linked to racial typologies, which associated place on the evolutionary ladder with racial biology (Radin 1929).

Anthropology, Boas and the culture concept

A short detour into the history of anthropology as a discipline is helpful in understanding changing concepts of race. Anthropology in the sense of the study of 'man' has been around for a long time (Hodgen 1964), but words such as 'ethnology' and 'ethnography' were not used in English until the first half of the 1800s, and anthropology as an institutionalised discipline only emerged during the nineteenth century, although with precedents in Germany stretching back to the late eighteenth century, under such names as *Volkskunde* and *Ethnologie* (Vermeulen 1995). In Great Britain the Ethnological Society of London was founded in 1842 and the Royal Anthropological Institute in 1871; in the United States the Bureau of American Ethnology was established in 1879; in France the Société ethnologique de Paris was founded in 1839 and the Musée d'ethnographie in 1878. Academic teaching programmes under the title of anthropology started in the universities of Oxford, Pennsylvania and Harvard in the 1880s.

Key figures in anthropological theory at this time – people such as Edward B. Tylor in Britain and Lewis Henry Morgan in the United States – were more interested in culture than race and biology per se. But they adhered broadly to social evolutionist ideas – with some variations – which placed 'primitive' people on a lower rung of the evolutionary ladder than more 'civilised' people. While they both showed some belief in the Enlightenment view that all humans possessed the capacity for reason, both suggested that the minds of 'primitive' people were structurally different from those of the 'civilised' (Stocking 1982: ch. 6). Late nineteenth-century anthropology thus supported contemporary racial theory.

It was against this background that Boas and others like him strove to separate biology (and race) from culture and to challenge social evolutionist approaches. For Boas the key point of anthropology was to explore the empirical cultural practices and products of so-called primitive peoples, rather than trace supposed social evolutionary progressions from promiscuity through polygamy to monogamy (Morgan) or from animism through polytheism to monotheism (Tylor). Race might exist as a purely biological reality, a realm of physical variation governed by Darwinian mechanisms, but this was a different realm from that of culture. For Tylor culture was a word applicable to 'primitive' peoples – hence his book *Primitive Culture* (1871) – which set him against those who used the word to designate the intellectual and artistic achievements of 'higher' civilisations. But culture was a single thing, a ladder or process of physical and mental transformation. For Boas the focus was on cultures, in the plural – the empirical diversity of actually existing peoples – and these cultures were separate from biology and race. The insignificance of race was also demonstrated for Boas by his finding, reported in *The Mind of Primitive Man* (1911), that, even at the level of biology 'the differences between different types of men are, on the whole, small as compared to the range of variation in each type' (cited in Stocking 1982: 192). This finding has been replicated many times since.

In short, Boas and others tried to radically separate biology and culture and confine questions of race to the realm of biology, such that race would only be about biology, seen as a physical substratum relatively unimportant in shaping culture.

4.3 Nazism, World War II and decolonisation

By the outbreak of WWII changes in scientific knowledge about human biology had already severely undermined the concept of race as a pure type and moved understandings of human diversity towards a population approach, which did not necessarily dispense with the terminology of race, but construed it as a way to talk about population diversity. If the term race already carried the taint of the scientific racism of earlier decades, Nazism charged the word with even greater negative meanings.

Nazi racial theories harked back to the hierarchies of scientific racial typology. They claimed to scientifically identify a master race, the Nordic race, which was the purest exemplar of the Aryan race, and a series of inferior or subhuman races, including the Slavs, 'Gypsies' (Romani), Jews and black people. As many Germans were not remotely 'Nordic', definitions were widened to include all Germans in the same superior category, but Nordic remained the ideal. These theories were disseminated through propaganda and legislated into bureaucratic procedures to determine race by criteria of Aryan or German descent. Those considered inferior were subject to forcible 'Germanisation', prohibitions against intermarriage, forced labour, sterilisation or physical elimination. The Jews were the primary target of these eugenic and racial hygiene policies, which culminated in mass extermination in Nazi concentration camps. Nazi ideas and practices were an important stimulant provoking worldwide rejection of scientific concepts of race and scientific justifications of racial inequality.

A second vital global shift set the scene for changes in ideas about race and theories of racial hierarchy. After WWII, European empires rapidly decolonised, pushed in most cases by unrest and armed resistance among colonised peoples:

- Indonesia was finally conceded independence by the Dutch in 1949, after prolonged unrest and several wars. Burma (1948) and Malaya (1957) won freedom from British rule, and Vietnam threw off French rule in 1954.
- In India, Gandhi had been leading protests since the 1930s, and the country, along with Pakistan, became independent in 1947; Ceylon (Sri Lanka) followed in 1948.
- In Africa, thirty-six countries achieved self-rule between 1944 and 1965.
- In the Middle East, Syria and Lebanon became free of French rule in 1945, while the British left Palestine in 1947, as Jewish–Arab violence mounted.
- The United States granted independence to the Philippines in 1946, retaining possession of many small territories (e.g. Puerto Rico, Guam, some Pacific islands).
- The Russian empire was different, as its territories became part of the Soviet Union from 1922 and countries such as Ukraine and Kazakhstan did not become formally independent until the 1990s, after the fall of the Berlin Wall (1989).
- The Japanese empire, which by WWII included large areas of China, South East Asia and many Pacific territories, collapsed after Japan's defeat in the war.

In most cases decolonisation involved non-European, non-white peoples throwing off the yoke of European white rule. Clearly, scientific theories that cast non-white peoples as racially, biologically inferior were hardly likely to be accepted by those seeking independence, membership of the newly formed United Nations (1945) and a seat at the table of world power. At the same time, European students of anthropology who carried out their fieldwork in colonial territories, while they did not by any means always challenge colonial rule, especially before the war, were increasingly critical of scientific racist theories (Conklin 2013).

4.4 UNESCO and after

Nazism was a key stimulus for growing critiques of racial science. Many social and natural scientists, even before the war, spoke out against Nazi racism. In 1938 the American Anthropological Association issued a denunciation of German anthropology in a statement that denied any connection between the biological variations of race and psychological or cultural qualities. Interestingly, the sister association of physical anthropologists did not issue such a statement; nor did anthropologists in England, where the Race and Culture Committee had advised against public pronouncements unless anthropologists could say positively what race was, as well as what it was not (Barkan 1992: 309, 339).

NEW TERMS FOR OLD CONCEPTS

In this context the terms 'ethnicity' and 'ethnic group' began to compete with race. The word ethnic is an old one, referring variously to differences of religion, nation and indeed race. It was often used by the speaker to label 'others' (e.g. pagans, non-English, non-whites).

One of the first uses of 'ethnic group' was in the book *We Europeans* (1935) by the biologist Julian Huxley and the anthropologist Alfred Haddon. This was an explicitly anti-racist text, written against Nazi theories. The authors recognised that Europeans differed biologically from other groups and among themselves, but rejected the use of 'race' to describe this, due to its association with racism and because it referred to a concept of race as pure type that could not be defended scientifically. They suggested the term ethnic group as a more neutral one, but it still referred to biological variation (Barkan 1992: 296–302).

In contrast, the concept of ethnicity was first used by the US sociologist and anthropologist W. Lloyd Warner, in a 1941 study of 'Yankee City', to capture social, not biological, diversity, which was seen as partly national (immigrant groups and their descendants) and partly racial (blacks and whites). The term race had been discredited by European fascism, so the old adjective ethnic was upgraded to an abstract noun. It hovered between being used only for 'others' in Yankee City – immigrants, blacks – and being used for everyone, including 'Yankees' (Sollors 1981).

Although the term race was falling out of favour, terms derived from 'ethnic' in fact retained race's ambiguous reference to both biology and culture.

After the war UNESCO convened a panel of experts – mostly social scientists – chaired by physical anthropologist Ashley Montagu, author of *Man's Most Dangerous Myth: The Fallacy of Race* (1942). An international body, an arm of the newly formed United Nations, was tasked with producing a globally authoritative statement about what was seen as a global problem. After consultation with biologists the panel issued a statement in 1950 that was meant to summarise in a nutshell the scientific consensus on race, clearly targeting the kind of theories the Nazis had touted. The statement (UNESCO 1952: 98–103) declared:

- Race should be understood, in biological terms, as a group of populations, which differed from each other in the frequency of one or more genes.
- Humans could be classified into 'the Mongoloid division, the Negroid division and the Caucasoid division'. (Note the avoidance of the term race.)
- Biological differences between populations, races or divisions were minimal compared to similarities and they could fluctuate over time.
- These biological variations had no correlation with mental abilities, cultural achievements or temperaments; they could not legitimise a hierarchy of races.
- Race mixture was not biologically harmful.
- As a result: 'For all practical social purposes "race" is not so much a biological phenomenon as a social myth' and 'it would be better to drop the term "race" altogether and speak of ethnic groups'.

As we will see, this statement caused some controversy at the time and was not accepted by all scientists, but the key point is that the panel of experts divorced biology from intelligence, culture and temperament. They recognised biological 'divisions' of human-kind but saw this biological variation as relatively minimal; in any case, given its history, the word race was not appropriate to describe this variation. Still, these experts used it themselves to refer to biological variation.

Race as a social construction

Since that time a great deal of work by geneticists and biological anthropologists has concluded that race has no biological validity at all. This means that race is *only* a 'social construction', that is, a way of thinking about human difference that has no grounding in any biological reality. The key arguments here are that, while biological variation certainly exists within the human species, this variation cannot be meaningfully packaged into anything that resembles the 'races' that earlier science supposed to exist or that popular usage still refers to.

HUMAN BIOLOGICAL VARIATION AND RACE

The idea of races as biological units is flawed because: (a) genetic variation is 'clinal', that is, the frequency with which a given genetic trait occurs in populations varies gradually across geographical space; and (b) the patterns of clinal variation for different genes are not 'concordant' (concordance means that where you find a given frequency of one trait you will find a similar frequency of some other trait).

Overall, the human species is a young one in evolutionary terms and did not evolve into geo-graphically located populations with a significant degree of genetic distinctiveness; the fact that there has been extensive movement and intermixture of populations has lessened still further the possibility of such distinctiveness. Thus, the genetic differences *within* categories commonly taken as 'races' are much greater than differences *between* such categories. Of the total biological variation found in samples taken from populations around the world, about 90–95% (depending on what type of variation is being examined) occurs within 'racial' categories and only 5–10% between them. Any set of people taken as a 'population' will display about 85% of all human genetic diversity (Brown and Armelagos 2001). These arguments have been made repeatedly in influential texts (Lewontin 1972; Marks 1995; Montagu 1942, 1964).

By the time the Human Genome Project sequenced a whole human genome in the early 2000s, a key finding – which has become a popular slogan – was that 'all humans share 99.9% of their DNA'. If all humans are biologically so alike, then the concept of race can only be a social construction: it has no basis in the actual biology of human diversity; instead, it is a way of thinking and talking about human differences that uses a varied mix of ideas about physical appearance, heredity, nature and culture, usually in relation to specific categories of people, labelled blacks, whites, browns, Asians, 'Indians', mestizos, Jews, etc. Biologists can statistically map the clinal distributions of the multiple facets of human biological variation, perhaps using the concept of 'population' to describe the geographical dimensions of that variation, while social scientists explore race as social processes and discourses. Saying race is a social construction does not mean it is not important: if people choose to discriminate (and even kill) on the grounds of what they think of as race, then it is extremely important. But it is also important to know that, if people justify their discriminations on the basis of real, natural, biological differences of race, there is no scientific basis for such differences.

WHAT IS A 'SOCIAL CONSTRUCTION'?

As this term is used so often to refer to the concept of race, it is worth thinking briefly about what it means. A social (or cultural) construction is anything that humans make. Sometimes it is assumed that a social construction is not real, it is just 'made up'. This is way off the mark. A house is a social construction, and also very real; its 'making up' requires a complex network of materials, people, social relationships, rules and ideas. We may think of a house being designed by an architect and built by builders, but these people emerge from a complex system of education about design and building, sets of norms and rules about professional architectural and building practice, and intricate systems that supply designs and building materials. Likewise, the social constructions that concern us emerge from very complex sets of interactive relationships and practices.

In the case of race, we are talking about a series of categories that humans use to classify others and shape behaviour towards them. All human classificatory systems are social constructions, but they use different criteria to organise the system.

Nations are social constructions that organise people into bounded territorial units, with political organisations, narratives of historical origin, symbols of identity and so on. These entities have a material existence (borders, armies, governments, flags, currencies), but they are all constructed and enacted by people. If a person kills another person, this may hardly appear to be a 'social construction', but if the killing is done in the name of the nation then it is exactly that. This is why nations are 'imagined communities' (Anderson 1983).

Economic classes are social constructions based on the criteria of wealth, control of resources, etc. Material things (land, machines, money) are involved, but they are all useless unless they are organised into social systems of production involving people in relationships. It may seem that, for example, keeping people off a piece of land by violence or the threat of violence is an irreducibly material act and not a social construction. But that violence has to be organised via social relationships – a family, a band of thugs, the police, the army – and very often it is legitimised by law, which is a social construction for the regulation of society.

Gender is a social construction based on the 'facts' of sexual difference. Different peoples around the world give different meanings to the categories 'man' and 'woman'. Indeed, the apparently irreducible 'facts' of sexual difference itself are (a) not so clear (are they about overall appearance, exterior genitalia, interior gonad anatomy, hormones, or chromosomes?); and (b) interpreted differently in different cultures, such that a simple physical distinction between 'male' and 'female' is not always present (Fausto-Sterling 2000; Herdt 1993). If violence is inflicted on women because of their 'natural' inferiority or difference, this is no less the result of a social construction than the violence visited upon people of other nations or other classes.

Race is therefore not unique in being a social construction, based on material facts (territory, wealth, sexual biology, human bio-geographical diversity), which in reality only draw their immense and often devastating power from the way humans selectively interpret them and organise them into specific configurations, which are always shaped by the functions they perform in the surrounding social context, a context usually structured by hierarchy and inequality.

used as evidence for arguments of this kind (Entine 2008). Others contend that social environmental factors push certain groups of people into certain sports, perhaps because racism restricts opportunities in other occupations (Malcolm 2012).

There is no doubt that certain geographically located populations or groups of populations may have certain genetic traits that, in the right social environment, allow them to compete very well at certain sports. But this has little relation to race, as it is popularly or even scientifically conceived. If it is proposed that 'blacks' as a biological race, for example, are genetically superior at sport, then we immediately have to recognise that very different genetics and bodies are required for sprinting (dominated by black Americans and Caribbeans, of West African and European descent) and long-distance running (dominated by East and North Africans). We also need to admit that West Africans themselves do not dominate sprinting, whereas the 'black' Australian world champion sprinter Cathy Freeman is genetically comparatively distant from many black American sprinters. So the notion of race, when construed in terms of 'black' and 'white', has little connection to the genetic components that might predispose towards good performance in specific sports (Malik 2009).

Another approach to sport, biology and race takes a different line, embracing biology but dismissing genetics. One argument is that blacks and whites in the United States have different experiences of learning basketball as they grow up. Blacks tend to compete on crowded inner-city courts and develop skills focused on dribbling, shooting under pressure, team work and so on. White men tend to play on less crowded suburban courts where they develop more individualistic skills. In the world of basketball the skills that black men tend to acquire are more successful (Harrison 1995). Black men and white men thus tend to develop different skills, which are actually embodied, that is, they become part of their body–mind habit structures. This is part of their biology, in the sense that habitual behaviour becomes part of the neurological patterning of the brain, but it is not part of their genes. This kind of approach is far removed from assertions about the genetic basis of race: it makes interesting links between race as a social force and biology as a flexible, developmental process; it tries to go beyond a simple divide between biology and culture and show how the two can interact (Goodman, Heath and Lindee 2003; Hartigan 2013; Wade 2002b).

4.8 Race, genomics and medicine: does race have a genetic basis?

The field of genomics emerged in the 1980s when technological advances in DNA sequencing made it possible to explore whole genomes of organisms – that is, their entire genetic make-up. Genetics had previously focused on very narrow segments of the genome or specific genes. In 2001–3 the Human Genome Project published the sequence of the entire genome of a human being (the geneticist Craig Venter) for the first time. A widely cited finding associated with these advances proclaimed that we are all 99.9% alike, a result that has been downsized by some scientists, including by Venter himself, to 99.5% or even 99.0%. These may seem tiny differences, but they are highly significant. Even if all humans are only 0.1% genetically different, that minute amount of difference must account for all the existing variety in the human species, visible or not.

HUMAN GENETIC VARIATION

DNA is made up of strings of four molecules called nucleotides, labelled A, C, G and T. The human genome contains about 3 billion of these DNA 'letters'. Although humans have nearly identical combinations of letters in their genomes, 0.1% of difference equals about 3 million possible places on the genome where a person can differ from another; 0.5% of difference would equal 15 million locations.

In practice, a key focus for the study of genomic variation is the SNP or single nucleotide polymorphism, which is a location on the DNA chain where just one DNA letter varies in at least 1% of the human population. Each SNP location can have up to four variants, one for each nucleotide. Estimates say there are between 10 and 30 million SNPs in humans; over 10 million have been identified so far. Many of these variants are located in so-called junk DNA, large segments of the genome that are not (yet) known to actively participate in guiding the formation of the proteins that make our bodies – although recent genomic science has been discovering functions that 'junk' segments of DNA fulfil in terms of regulating gene activity. This aside, there are millions of significant locations on the genome where one human differs from another.

Geneticists have been exploring in detail the small areas of variation in the human genome and have come to different conclusions about how to conceptualise genomic diversity and its relationship to race. On the one hand, the finding that all humans are 99.9% (or perhaps only 99.5%) the same seems to put the final nail in the coffin of the concept of race, which implies important biological differences between different types of people. On the other hand, within that 0.1% (or 0.5%) of difference, some geneticists maintain that variation is structured along lines that correspond, more or less, to the racial populations of older science and popular culture. As populations evolved in particular niches, with gene flows restricted by geography and culture, they developed distinctive genetic profiles, identifiable in terms of the relative frequency of genetic variants, or sometimes in terms of the simple presence of specific mutations, which spread within a given population. Some of these populations, according to some geneticists, are continental bio-geographical entities that correspond to well-known categories such as Africans, Europeans, etc., or, if we include their migratory descendants, to 'blacks', 'whites' and so on.

The jury is out on this question: some argue that 'racial and ethnic groups do differ from each other genetically' and these differences have 'biologic implications', including susceptibility to disease (Burchard et al. 2003). Others reply that '[genetic] variation is continuous [across the globe] and discordant with race, systematic variation according to continent is very limited, and there is no evidence that the units of interest for medical genetics correspond to what we call race' (Cooper, Kaufman and Ward 2003).

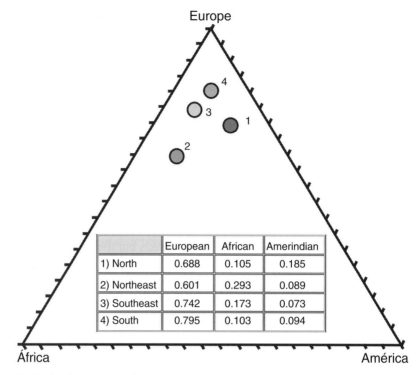

Figure 4.1. Graphic from Pena et al. 2011

(see Fig. 4.1) displays a triangle, the three corners of which are labelled Europe (at the top), Africa and America; inside the triangle, Brazilian samples are placed to indicate the relative amounts of each ancestry they have in their DNA (Pena et al. 2011). In these and other such studies it is common to find statistics for individuals and populations that specify the proportion of their genetic ancestry that comes from African, European and Amerindian origins (Bedoya et al. 2006).

These graphics and figures give the impression that Europeans, Africans and Amerindians are discrete populations, genetically very different from each other. If all humans are 99.5% the same, what exactly does it mean to say that, for example, 33% of a person's genetic ancestry is of European origin? The genetic material from Europe will be 99.5% the same as the genetic material from Africa or native America, but the dividing up of a person's ancestry by continental origin hides that commonality. Now, the word race is never mentioned here (indeed, the geneticists say it has no biological validity), yet the populations named are very familiar in terms of long-standing racial classifications – and they seem to have a genetic basis.

If race seems to be reproduced in genetic idiom in these genomic projects, we need to be clear here about exactly what is happening. For the geneticists the issue of race does not arise, because they are referring to genetic ancestry. They are talking about a relatively small number of genetic variants that are statistically more common in, or occasionally more or less exclusive to, certain parts of the world (e.g. West Africa). They are not talking about the overall biological profile of a population. Thus when they say that 33 per cent of a person's genetic ancestry comes from Europe, they are basing this on the analysis of, say, forty specific markers, which have been specifically chosen to differentiate as much as possible between African, European and Amerindian samples. This is then extrapolated to make an estimation about the geographical origins of a person's ancestors. Also, the genetic variants employed are often ones that reside in the 'junk' DNA and thus have no (known) influence on a person's phenotype. Finally, the HapMap Consortium insists that a sample from, say, Utah should not be generalised to represent 'Europe': it is just a sample of some people from Utah (International HapMap Consortium 2003).

However, there are also a number of countervailing tendencies. First, as noted, the simple reiteration of difference tends to hide commonality and represent specific populations as genetically distinct. Second, it is clear that geneticists do, in effect, generalise geographically from the specific population to a whole continent. 'Yoruba from Ibadan' are made to stand in for 'African ancestry' (Bliss 2009). Third, there is a historical generalisation being made: contemporary Zapotec residents are being made to stand in for native Americans from five hundred years ago. Any genetic changes over that time are being ignored, for the purposes of the genomic study in hand, and 'Amerindians' are being represented as genetically homogeneous over time and space. Fourth, it is common in this kind of sampling process to use individuals who are genealogically rooted in the local population. Thus the Zapotec people sampled had to have grandparents who spoke the local language. We can easily see why geneticists do this: they want to sample people who are likely to be good representatives of ancient Amerindian genetic ancestry, not some recent migrants to the community who, although accepted as Zapotec by local standards, may have a mixed ancestral heritage. But the effect is to 'purify' the Zapotec sample so it looks genetically more homogeneous (Reardon 2008). These tendencies reinforce the way populations become racialised and geneticised.

Who is doing the racialisation?

If populations are becoming racialised in genomic research, two problems arise in relation to who is doing what in this scenario. First, the way the data are displayed 'seems' to suggest familiar notions of race. But *to whom* does it suggest this? Geneticists know they are talking about genetic ancestry, which is not the same as biological race. So the problem then lies with other people, less expert in genetics, who interpret the scientific findings, perhaps via the media, where some of these findings are often presented, or perhaps via the more popular writings of the geneticists themselves, some of whom disseminate their research in accessible books, newspapers and websites. When less expert people in Latin America see results about proportions of ancestry, they may easily assimilate such findings to common-sense ideas about *las tres razas* (the three races) that are popularly said to have formed Latin American nations and which, in many nations, can still be found today as *los negros* (the blacks), *los indios* (indigenous peoples) and *los blancos* (the whites), whose mixtures are evident as *los mestizos* (the mestizos).

A similar focus on the nation can be seen in a genomic project called People of the British Isles. Like many other projects, this sampled people in Britain, or rather – and this is vital – those whose parents and grandparents had also been born in the 'rural' locality where they were born (Winney et al. 2012). The objective was to collect data on possible genetic predispositions for certain disorders and to trace old patterns of migration in the region, including Anglo-Saxon, Viking and Norman invasions. The picture produced of the British Isles is far removed from the idea of a pure, ancient British stock. On the contrary, it is more akin to Daniel Defoe's description, in his satirical poem 'The True Born Englishman' (1701), of the English as a 'mongrel half-bred race', made up of bits of ancient Briton, Pict, Scot, Roman, Saxon and Dane. Defoe's aim was to ridicule English xenophobia, based on ideas of racial purity, but this was not the objective of the People of the British Isles project: a picture of the British as ancestrally diverse emerges, with no reference to race at all, or even ethnicity. However, racialised meanings enter via the frame of nation. First, as with other projects of this kind, mixture implies original ingredients: 'Englishness is not a genetic category … but other ancestries are presented as if they correspond to genetically and geographically bounded groups – Sub-Saharan African, Native American, Northern European' (Nash 2013: 201).

Second, by focusing on rural people with locally born grandparents and also on ancient mixtures, most of the more recent and non-white immigration from Britain's ex-colonies is simply ignored, thus implicitly representing the nation as white (Nash 2013). Indeed, commentators on the project's first scientific paper noted that the medical use of the project's data would be problematic as many cases of disorders would come from 'multi-ethnic cities' (Tyler-Smith and Xue 2012).

FORENSIC GENETICS AND RACE

Forensic genetics seeks to identify individual people or bodies using their DNA. This usually occurs in criminal investigations where DNA traces have been left at a crime scene and investigators want to link the traces to a suspect or to use them to generate possible suspects; or in attempts to link living relatives to dead bodies, found at crime scenes, in mass graves, etc.

One way that race can enter into forensic genetics practice, in a way suggesting that race has a biological reality, is in the use of DNA traces to generate racialised descriptions of possible suspects. For example, in 2003 Louisiana police were tracking a serial killer they believed was white; a private DNA testing company, DNAPrint, analysed a DNA sample and concluded that the killer was likely to be a light-skinned black man (in terms of US social categories), as he had genetic markers indicating significant African ancestry. It turned out the killer was black. In Britain, too, government forensic scientists claimed in 1993 to be able to distinguish between 'Caucasians' and 'Afro-Caribbeans' in nearly 85 per cent of cases, using specific combinations of genetic markers.

These are neat examples – and they suggest that there is a biological reality to common-sense categories of race as they are deployed in everyday life – but their neatness is misleading. Such profiling can be a useful guide in some cases, when a local population is made up of two or three groups with recent ancestors from geographically distinct areas of the world (such as Britain and the Caribbean), but it can only provide an indication and a probability. Precisely because the genetic markers used are usually not related to phenotype, a person with considerable European ancestry may not 'look' white (Ossorio and Duster 2005; Sankar 2012). This is evident in the case of Neguinho, described above. In addition, suggestions drawn from DNA data will always be assimilated to existing social categories and ideas. Thus in Holland, if DNA data suggested that a suspect was 'of Mediterranean ancestry', it is likely that would be interpreted by many Dutch people to mean 'of North African appearance', given the presence of North African immigrants and prejudices about them and their supposed criminality (M'charek 2008).

CONCLUSION

In this chapter we traced moves away from the racial typological science of the late nineteenth and early twentieth centuries. Fundamental to this was a shift away from the (almost) fixed and permanent type towards the concept of dynamic and changeable populations. Also vital was the attempt to divorce biology from culture and consign the former to a lesser or negligible role in explaining the latter. A forceful push to these processes was given by Nazism and WWII, which introduced a globalising anti-racist agenda that set a new context for the concept of race – one which began to brand the concept as toxic and to place emphasis on culture as the way to conceive of human diversity.

The shift away from race and towards culture, however, is complex and contradictory. We have seen that the questioning of race even as a biological reality has been uneven: we have seen the persistence of theories that hold that (a) humans can be divided up into distinctive biological units, similar or equivalent to the units that used to be called races; and (b) these differences may be reflected in important dimensions such as intelligence, genetic profiles relevant to medical well-being and illness, and even identity.

This qualifies the standard idea that, since WWII, there has been an overall and radical shift to (a) the idea of race as a social construction, with no basis in biology; and (b) a cultural form of racial thinking, in which biology disappears from the modern concept of race (or at least takes a very diminished role) and difference is conceived only in terms of culture. The evidence presented here indicates that race and biology are still linked in some areas of the life sciences and that, even though the term race is generally not used, it is often implicit and may even be an explicit referent. In the next chapter I take this discussion to another level, that of institutional state practices and everyday life.

DO WE NATURALLY PERCEIVE RACIAL DIFFERENCE?

As a segue into the next chapter, I want to present a different kind of argument that roots race in human nature, but is completely different from the approaches looked at above which argue for a biological basis for racial differences among humans. This argument proposes that humans are naturally predisposed, for evolutionary reasons, to (a) sort people into categories; (b) attribute them with essential, typical characteristics; and (c) act on those differences, notably in the direction of 'sticking with one's own', that is, with people perceived to be like oneself. The categories and the criteria for defining them can be various: they could be physical appearance or cultural traits. The point is that the theory supposes that humans have been genetically wired by evolutionary processes to form in-groups and distinguish them from outsiders, whether to keep them at arm's length, fear them, or exchange with them. In these theories the supposition is that in-groups tend to be defensive and keep outsiders at bay; this is held to explain racism, xenophobia, ethnocentrism and so on (van den Berghe 1979). As we will see in the next chapter, such a theory is not just an academic proposition, but is a popular belief that human cultural groups are fundamentally different and it is best 'to stick to your own kind'; this is 'cultural fundamentalism' (Stolcke 1995).

One sophisticated version of this argument pertains specifically to the perception of racial differences, rather than cultural ones, and thus belongs in this chapter, because the argument says that we are naturally predisposed to see not just difference, but 'racial' difference. Basing himself in psychology and cognitive anthropology, Hirschfeld (1996) argues that people think about people in a particular way, dividing them into 'enduring types', perceived to have 'clusters of naturally grounded properties'. This Hirschfeld calls a race theory or concept, and he sees it as a universal feature of the human mind. This basic cognitive mechanism develops into specific forms of 'racial thinking' in particular historical and social circumstances – these include English racial thinking about the Irish, European ideas about Jews, colonial perceptions of races, etc.

Hirschfeld supports his argument with data from psychological experiments with young children in which he tests how the children view certain kinds of visible features, not just phenotype, but also clothing and markers of occupation (e.g. a doctor's stethoscope). He tests which features are seen as (a) the most salient; (b) the most likely to be transmitted to offspring; and (c) the most likely to define a person's behaviour. The data show that 'racial' features, such as skin colour, score highly in children's perceptions. As his subjects include pre-school children, this is evidence for Hirschfeld that these patterns are universal traits, not the product of social influences.

A key problem with these experiments is that they are almost all carried out with children in the United States, with just one sample from France. The data could just as easily show that even young children in these countries pick up the importance of racial differences in their social surroundings. While humans may well share certain basic cognitive capacities, it is extremely uncertain exactly what these might be, and how specific and predefined they are. Hirschfeld's data give no basis for positing a universal mechanism that is preconstituted to perceive the kinds of physical difference that have come to be called 'race' in the history of Western thought and colonialism.

FURTHER ACTIVITIES

1. Look at UNESCO statements on race and compare them (e.g. at http://www.honestthinking.org/en/unesco/index.html). See UNESCO's discussion of the 1950 and 1951 statements in the document *The race concept: results of an inquiry* (UNESCO 1952) (see http://unesdoc.unesco.org/images/0007/000733/073351eo.pdf).

2. Search the internet for discussions of the idea that 'we are all 99.9% genetically the same'. You will see that there is a good deal of debate about this. What is at issue? How does it relate to race?

3. Seek out some of the writings about race and sport, starting with Entine (2001, 2008) and going on to critiques of his work (Malcolm 2012; Malik 2009). You might want to find some reviews of Entine's book. Does Entine have a leg to stand on?

4. Hirschfeld's book on the perception of race (Hirschfeld 1996) was the subject of a series of commentaries in the journal *Ethos*, volume 25, issue 1, 1997. Have a look at the critiques of his ideas and his response to them.

5 Race in the era of cultural racism: politics and the everyday

5.1 Introduction

In the preceding chapter we explored how the concept of race as a biological reality has persisted in a number of ways in the spheres of biological science, psychology, medicine and forensics. In this chapter I look at how race as a concept and racialised ideas about nature and biology appear (and disappear) in the domains of politics and the everyday. This is potentially a vast topic, and more detail about it will emerge in subsequent chapters that explore how race operates in the social and political life of specific regional contexts. What I am interested in in this chapter is seeing how a more or less explicit reference to 'race' remains part of public institutional practice in an era of cultural discourse and how a concern with race as phenomenon involving biology, heredity and nature – not just culture – also persists in the public and everyday domains.

Of science and society

A caveat: the fact of discussing first 'science' and then 'politics and the everyday' should not be taken to imply a separation between science and society, because science is itself a social activity and is deeply implicated in social policy. The findings of science about race and medicine have vital practical and policy implications (e.g. about how best to practise diagnosis, how to design screening programmes, how to address public health problems such as heart disease or hypertension), as well as shaping the economic workings of pharmaceutical companies which may choose to target 'ethnic' minorities with specific treatments. What genomics has to say about human diversity and race in forensics has important consequences for policing, the collection and storing of DNA in databases and the possible targeting of racial minorities in the criminal justice system. If we were to believe the race and IQ theorists, this would have radical consequences for education policy.

More broadly, science is deeply embedded in a social context, even while its internal problem-solving is not determined by this context: as we saw in the last chapter, it is hard to draw very tight connections between specific racial theories in late nineteenth-century science and specific configurations of empire; likewise, the demise of scientific racism started while European empires were still at their peak. But it is also impossible to ignore the broad correlations between empire, abolition and the rise of scientific racism;

and between decolonisation and anti-Nazism, and the strengthening of anti-racist and anti-racial thinking.

How to recognise race in an era of culture

In thinking about how race as a term and a concept appears and disappears in an era of cultural discourse, we need to look at two things. First, we can trace the explicit use of the words race and racial, which by no means entirely disappear, even though they may be considered tainted by the invidious ideas and practices of racism (a word which remains common usage).

Second, we need to be alert to instances where, although the word race does not appear, the race concept is expressed with a different form of words. How do we know when this is happening? What identifies something as being racial, when it is not named as such? This may not be a problem in some contexts. When football fans in Europe direct monkey chants or throw bananas at black players – very recent examples of which have been recorded in Russia – or when Italy's first black African-born minister, Cécile Kyenge, likewise had bananas thrown at her in public in 2013, probably by right-wing activists, we can easily recognise these as racial instances. By implicitly comparing black people to monkeys, the banana imagery clearly invokes racial ideologies, even if the right-wing criticism of Kyenge is publicly expressed in terms of opposition to her immigration policy proposals, which facilitate citizenship for immigrants. In more institutional or official contexts, it is usually less easy to encounter such overt use of racialised discourse, although during a political rally an Italian right-wing parliamentarian did compare Kyenge to an 'orangutan', and football commentators in Britain have occasionally been caught making overtly racist remarks when they thought they were off-air.

Such clear instances aside, racial thinking is often not so easily identified. In accordance with the approach to race outlined in Chapter 1, my view is that we can identify unnamed race by the presence of two key features: (a) the categories being talked about are the same as or very similar to the kinds of categories found in an explicitly racial discourse, usually those that have emerged in the history of Western modernity and colonialism and in associated elaborations around race: black, white, Asian, indigenous, etc.; (b) the way these categories are defined and the way individuals are allocated or allocate themselves to them depends at some level on ideas about the heritable bodily differences associated with such historical categories – differences understood as linked to cultural differences, albeit not necessarily in a deterministic and immutable fashion. These two features need not always both be clearly present at the same time: one may connote the other; talk of 'blacks' and 'whites', for example, usually connotes ideas of bodies, descent and heredity.

Criterion (b) is particularly tricky to detect in the domain of politics, policy and public institutions because the vast majority of identifications made in this domain are *self*-identifications. In a census, for example, the blacks, whites and indigenous peoples are all those who choose to identify as such: this immediately consigns to the background the basis on which someone makes such an identification. Are they focusing on their skin colour, other aspects of appearance, who their parents are, what other people call them, where they live, how wealthy they are or what their political views are? We don't know – unless we do some ethnography and explore what drives people's identifications. In the public institutional domain, however, it becomes quite easy to disavow any idea of race. So we have to look beneath the surface and try to ascertain whether a concept of race is at work.

Luckily, in this institutional domain we are often handed rather a big clue: 'ethnicity'. This is often the term of choice in institutional discourse and practice to refer to differences that used to be called racial. As noted earlier, although it is generally deployed today to refer only to 'cultural' difference, the term ethnic actually retains in its roots an ambiguous duality, evoking both biology and culture. Thus, while ethnicity might easily be used to refer to the differences between, say, Polish and English people in London, it is very commonly used as a simple euphemism for race. So when we see 'ethnicity', we know to keep an eye out for 'race'.

5.2 The institutional presence of race

The first thing to note about the apparent replacement of 'race' by 'ethnicity' or 'culture' in the spheres of politics and public institutions is that this effect is rather uneven. In France or Germany, for example, a combination of post-Nazi politics plus a determined adherence to principles of liberalism has meant that the word race is rarely found. In 2013 the French National Assembly voted to remove the word from all state legislation. In social science circles in these countries and many other European countries there is traditionally a deeply rooted suspicion about use of race as an analytic concept to describe certain modes of thought or practice. The term itself is tainted and its use almost taboo: to deploy it carries the threat of seeming racist oneself by giving currency to a concept so often used in the service of racism. As noted previously, some people in the United States prefer to avoid explicit use of the term and to avoid explaining things in terms of racial difference, in case this is interpreted by others as racist (Bonilla-Silva 2003; Frankenberg 1993). In all these cases the word racism is more easily found, used to describe a deplorable and erroneous belief and practice, usually associated with people other than oneself.

Yet in some contexts the term and the concept of race have a powerful institutional life, even if they are sometimes euphemised as ethnicity (or national origin). Often this life is rooted in a history of slaveholding and racial domination, and is nowadays related to priorities of social equality, inclusion and anti-racism. As we saw in Chapter 1, the easiest place to find the word race or racial in institutional domains worldwide today is in *anti-racist* legislation or in constitutional clauses that *prohibit discrimination* on 'grounds of religion, race, caste, sex or place of birth' (this text comes from the Constitution of India, clause 15, but similar wording can be found in many others) (Bonnett 2000).

Enumerating by race: towards 'inclusion' and multiculturalism

A study of ethnic enumeration in the year 2000 censuses of 141 countries found that 63% of them included some form of 'ethnic' counting. Of these, about 15% actually used the word race, often alongside another term (colour, ethnicity), while many more used the word ethnicity (56%) or nationality (23%); another 15%, mainly in Latin America, but also the United States, Canada and Australia, asked about membership of an indigenous group or tribe. The censuses that used the word race were either in the Americas (e.g. United States, Anguilla, Bermuda, Brazil, Jamaica, Saint Lucia) or were United States territories (Morning 2008).

Sometimes, although the word race is not used, the classification underlying the census is clearly racialised. Many census questions about indigenous group membership – for example, in Latin American countries – effectively separated indigenous from non-indigenous people, which is a racialised divide. After 1976 the Australian census dropped a question about 'racial origin' (options were European origin, Aboriginal or Torres Strait Islander origin, or other) and asked instead about 'Aboriginal origin', but the question still clearly separates black Aborigines from all others. The Colombian census of 2005 has various options in its ethnic question, but the results are aggregated to divide the population into Afro-Colombians, indigenous people and, by implication, neither of the above. The British census has many options for 'ethnicity', but the major categories are White, Asian, Black, mixed and other. The 'population groups' of the South African census are the same as the apartheid-era races. The countries that have either an explicit race question or a clearly racialised ethnic question tend to be those with a history of slavery and/ or racial oppression.

The reasons for ethnic enumeration, while they were once related to colonial governance, are nowadays usually related to liberal governance. Counting people has long been a fundamental tool of governance: technologies of counting pull people into the web of state practices, while the data generated give the state a periodical cross-section of the nation-state and provide the basis for policy making. Enumeration, statistical fact making and precision are all important ways for states to know about society and to organise and govern people (Poovey 1998; Wise 1997).

The dominant idea behind counting ethnic and racial difference is to map the dimensions of racial and ethnic inequality, with the ultimate aim of correcting these (although this is not the only aim, as we will see). South Africa used to have strict racial counting procedures as part of the post-1948 apartheid regime in which the 1950 Population Registration Act required every citizen to be classified and registered according to a racial category – White, Black, Coloured and (later) Indian – that largely defined civil rights. Post-apartheid counting of 'population groups' aims to monitor racial equality, not to enforce racial inequality.

The Brazilian census has included a colour question since 1872 (Nobles 2000): data have been used to document racial inequality and to demonstrate that some of it is caused by current racial discrimination, rather than just being a hold-over from the days of slavery (which only ended in 1888) or being the result of race-blind class discrimination. From the 1990s the state has embraced measures aimed at correcting racial inequality, with race-based affirmative actions, including (as we saw in Chapter 4) racial quotas for many public universities, a Secretariat for Policies Promoting Racial Equality (SEPPIR, founded in 2003) and a 2010 Statute of Racial Inequality. For Brazil it is clear that the continued institutional vitality of the term and concept of race is due in great part to the growth of an anti-racist agenda.

This shift towards enumeration for anti-racist objectives is also obvious in the United States and Britain, where from the 1990s the state has mandated the collection of gender and race data in clinical research and/or practice. Policy measures such as this obey a general trend towards the 'inclusion' of diversity in medical research, which is partly the result of feminist and minority advocacy (Epstein 2007). Canada has embraced a similar convergence between official multiculturalism and the inclusion of ethnic diversity in the practice of medical research (Hinterberger 2012).

RACIAL CATEGORIES IN MEDICAL PRACTICE: TO USE OR NOT TO USE?

There are heated debates about collecting data on race in clinical contexts. On the one hand, it allows the mapping of racial disparities in access to health services and in health outcomes. Such data could be used to make an argument that disparities in health outcomes (e.g. higher levels of heart disease in racial minorities) are due to racial inequalities in the society more widely (such as higher levels of unemployment and poverty, poorer living conditions).

On the other hand, critics argue that the official and continuous use of such categories legitimises and reinforces their use in the wider society, which can result in the strengthening of racial barriers, greater segregation and the hardening of racially prejudicial attitudes.

Just as important, the use of racial categories in medical practice invites both researchers and people more generally to make biological connections between social categories of race and certain health conditions or outcomes. While some scientists think such genetic connections may exist (see Chapter 4), many of them think that social categories of race or ethnicity are rather poor proxies for the detailed data about individual genetic ancestry that might be useful in making a diagnosis or prescribing treatment. In this view, then, the use of social categories of race, while potentially useful as a source of standardised data about racial inequality, also brings significant risks (Ellison et al. 2007; Smart 2005; Taussig and Gibbon 2013).

These moves towards inclusion in medical institutions are part of a wider trend towards state multiculturalism, which is evident in many liberal democracies. In the wake of decolonisation, increased post-colonial migration at a global scale, the erosion of Euro-American economic and political dominance, and the burgeoning of social and philosophical movements celebrating diversity of all kinds and challenging Western superiority, racism, sexism and heterosexism, the official recognition of diversity has become part of the definition of a modern liberal democracy. Multiculturalism involves naming and cataloguing diversity, and, in some contexts – typically those marked by histories of slavery and racial oppression – it becomes a key vehicle for talking about racial difference, without necessarily mentioning the word; as usual 'ethnicity' or 'culture' are convenient euphemisms.

Although the official recognition of diversity is uneven between countries and within countries in terms of the specific domains in which it operates (Queen's University 2014), a general move towards multiculturalism can be seen in many institutional fields and in many countries. The state and other public bodies make provision to recognise ethnic diversity, with the aim not just of celebrating it, but ensuring equality, non-discrimination and, especially, anti-racism. In this context 'racism' is often conflated with many kinds of discrimination on the basis of 'difference' – ethnic, national, cultural, religious – but racial difference remains one of the key, albeit unspoken, targets of multiculturalism. Health is one field for inclusive policy where race appears, as noted above; another major one is education, given its perceived importance in defining people's life opportunities in society and in forming citizens. Multiculturalist educational policies include providing special opportunities for minority groups (such as the university quotas in Brazil), giving access to bilingual education, adjusting educational curricula to reflect minority group histories and cultures, and training educators to be sensitive to cultural differences in the classroom.

Like the debate about the use of racial categories in medical practice, multiculturalism, especially when endorsed and practised by the state, is often seen to be a double-edged tool (Lentin and Titley 2011; Modood 2007; Taylor and Gutmann 1994). On the one hand, public spaces are created for groups that were formerly sidelined and marginalised. Valuable economic and cultural resources may be channelled towards them. The notion of multiculturalism bespeaks tolerance, equality and liberty – all key liberal values (see also Chapter 8).

On the other hand, group identities may be reinforced, resulting in the growing fragmentation of society into many diverse interest groups and the undermining of solidarities built around common agendas (of, say, poverty, employment, environment, women's rights, etc.). Multiculturalism may mean the packaging up of citizens into cultural ghettos, bringing division, lack of communication and intolerance. Some see multiculturalist inclusion as itself a form of control, a specific variant of liberal governance, in which difference – which, if repressed, could become troublesome – is recognised and fostered, *within certain limits*, defined by the state (Hale 2005). The state may actually find some advantage in dividing up society into many groups, incorporating them on its own terms and putting them into competition with each other for resources and attention: there is a certain 'cunning' behind the politics of recognition (Povinelli 2002).

For our purposes the point is that, like inclusive agendas in medicine, multiculturalism is an institutional domain where racial difference is addressed. The concept of race may often be explicitly named only in relation to anti-racism, but the differences seen as relevant to multiculturalist policies are, in those contexts marked by histories of slavery and racial oppression, clearly racial ones (Lentin 2004).

Enumerating by race: control and exclusion

As one might expect, not all instances of the institutional presence of race or race-like versions of ethnicity in official domains today are related to anti-racism and liberal multiculturalist versions of inclusion. This is not surprising, given the historical baggage of the term and the role racial difference and racial inequality continue to play in everyday life. If the public institutional view is that the word race is best mentioned only when prohibiting racism, the fact remains that racism and racialised identities continue to be a very salient aspect of social life in many countries, even if almost no one admits to actually being a racist (even those widely labelled as racists, such as the English Defence

League or the Ku Klux Klan). It would be strange if this were not reflected in the practices of public institutions.

Yet we are on difficult ground here, because it would, after all, be a blatant contradiction of post-war anti-racist liberal principles – according to which race is usually named only in order to disallow racism or to facilitate measures for racial equality – to legislate explicitly in favour of *racial* or even ethnic exclusions. Certain elements of public opinion might be explicit about race and ethnicity – for example, in everyday talk about the control of crime or immigration – but institutional policy must be seen to be non-racist, even if policy makers and in particular politicians are responding to pressures from different segments of the voting public. Therefore institutional policy generally avoids all mention of race and ethnicity in relation to exclusion and discipline. Under these circumstances, we need to explore how certain policy measures effectively target racialised categories. The issue of whether this is done intentionally by policy makers is always a murky one, but in some cases the evidence is fairly substantial.

Exploring how race appears in institutional practices of control and exclusion in effect opens up the entire area of the operation of racism – in employment, housing, policing, education, etc. – which we will explore for specific regional contexts in later chapters. For now I will sketch two key disciplinary domains of institutional practice where race (or ethnicity, used as a euphemism for race) enters as an implicit factor in processes of control and exclusion.

The criminal justice system. In Britain the criminal justice system records and monitors ethnic and indeed 'racial' data at various stages of its operation, from the people the police stop in the street to the way sentences are handed out. Britain's Ministry of Justice produces annual statistics on 'Race and the Criminal Justice System' to 'assess whether any discrimination exists in how the CJS treats people based on their race' (Ministry of Justice 2011). This obeys the aim of anti-racist inclusion, but the measure is also a response to the common perception of a deeply embedded racism in the police force, going beyond the scarcity of non-white police officers and the possibly racist attitudes of individuals within the police force, to encompass the practices of the force as a whole. The police began in the 1970s to note the racial identity of people they arrested and used the data to make statements about the supposed propensity of young black men to commit crimes such as mugging; these data legitimised police tactics of targeting young black men for stop-and-search procedures. If black people were stopped disproportionately often, incidents of law-breaking found among them would also be disproportionately high (disregarding the possibility that infractions might be falsely pinned on suspects). This creates a vicious circle of suspicion and confirmation of suspicion that criminalises young black men and, in Britain, led to a moral panic around race and crime (Hall et al. 1978).

A key event in the move towards the use of racial data to monitor racial discrimination in the criminal justice system was the murder of black teenager Stephen Lawrence in 1993, for which no one was convicted until 2012, after lengthy campaigns by his family and an inquiry that concluded that London's Metropolitan Police Force was 'institutionally racist'. In the words of the inquiry report, this meant not that all police officers were intentionally racist (although some were) or that the force had racist policies, but that, for example, the police had initially failed 'to recognise and accept racism and race relations as a central feature of their investigation' and that a culture existed among officers in which, based partly on their own experience in policing, many of them stereotyped black people as likely suspects (Macpherson 1999: ch. 6).

Similar patterns of 'racial profiling' – targeting people of a particular racial category as potential suspects, usually on the basis of their appearance alone – can be found in Australia, where Aboriginal people tell of police harassment (Cowlishaw 2008). The British example also finds resonant echoes in the United States (see Fig. 5.1), encapsulated in the ironic reference to people stopped for DWB or 'driving while black' (Ossorio and Duster 2005) and in high-profile incidents such as the videotaped police beating of black construction worker Rodney King in 1991, for which no officers were initially found guilty.

Referring back to the section on forensic genetics in Chapter 4, some critics worry that the marked racial disparities in the US and British criminal justice systems will lead to black people being overrepresented in the rapidly growing databases that store the DNA of people who have been convicted or even just arrested (Duster 2006). For example, in Britain 37% of black men have their DNA profile on the police database, compared with 13% of Asian men and 9% of white men. This statistic is known because in Britain the 'ethnic appearance' of each person is noted in the database (Randerson 2006). Critics argue that white people will be less likely to be caught in DNA database searches, because their criminal activity is less likely to have been detected and their DNA included in the database.

Figure 5.1. Plainclothes police detectives searching suspects in Harlem, New York City
(© Michael Matthews – Police Images/Alamy)

Immigration. Historically many states have used immigration control to manage the racial composition of their populations, with a view to producing a nation deemed not just governable, but morally good. In Europe and the Americas the explicitly racial exclusions that were common for much of the nineteenth and twentieth centuries became politically difficult in the increasingly anti-racist climate of the post-WWII world, yet many states remained concerned about non-white immigration (FitzGerald and Cook-Martín 2014; Schain 2012).

The United States virtually banned Chinese immigration in the 1880s, later extending this in the twentieth century to exclude immigrants of Asian descent. From 1924 to 1965 a national quota system for immigrants was used to give preference to Northern and Western European migrants and restricting Southern and Eastern Europeans, seen pre-WWII as not – or at least less – white, and less desirable in eugenic terms. Of course, immigrants from outside Europe were even more restricted. Mexicans were admitted as temporary labour, for example in the Bracero programme (1942–64), on the assumption that they would return home, which they were occasionally forced to do, as in the 1954 Operation Wetback that rounded up and deported many Mexicans. As late as 1960, about 75 per cent of immigrants to the United States came from Europe.

The United States was actually quite tardy in the Americas in lifting explicit racial restrictions on immigration in 1965 with amendments to the Immigration and Nationality Act: other less democratic nations in Latin America, such as Chile, Uruguay, Paraguay and Cuba, had done so some decades earlier (FitzGerald and Cook-Martín 2014). The immigrant population increased, doubling from about 10 million (5% of the US population) in 1970 to 20 million (8%) in 1990 and 40 million (13%) in 2012. Latin American and Asian migrants made up much of this increase: by 1980 they outweighed European immigrants for the first time, and by 2012 constituted over 80% of the total immigrant population (Migration Policy Institute 2014).

The United States was not alone in the Americas in wanting European migrants pre-WWII. Brazilian politicians were keen on attracting European migrants, and some of them proposed racial immigration bars; in the 1920s Brazilian consuls systematically refused visas to US black applicants. As late as 1945 the authoritarian government of Getúlio Vargas, before its fall later that year, passed a decree which said that immigrants should be admitted bearing in mind the need to 'preserve and develop, in the ethnic composition of the population, the more desirable characteristics of its European ancestry' (Skidmore 1990: 25). Venezuela enforced a ban on non-white immigration until 1945 (Wright 1990: 105).

Australia's immigration laws from the 1850s until WWII did not always explicitly exclude racial groups – instead they prohibited people with manual labour contracts and imposed a written test in English – but policy was consciously designed to promote a 'White Australia' (see Fig. 5.2). Restrictions were eased from 1949, but not finally dismantled until 1973. Since then, non-white immigration – mainly from Asia and the Pacific – is often seen as a problem and targeted by politicians who claim to be defending Australian citizens' welfare against 'boat people'; meanwhile white immigration – usually arriving by air – is of less public concern, even though immigrants from Europe and New Zealand easily outnumber those from Asia, Africa and the rest of Oceania combined (Cottle and Bolger 2009; Hage 2000; MacCallum 2002; Tavan 2005). In 2014 the government decreed that

Figure 5.2. Article from Australian newspaper the *Daily Mirror* supporting the White Australia policy, 11 March 1954.
(National Archives of Australia: A1838, 581/1 Part 2)

asylum-seekers who, since 2001, had been temporarily placed in camps in Papua New Guinea would, even if they eventually won refugee claims, be resettled in Papua New Guinea but never in Australia. Although Australia's asylum-seekers come from many areas, many of them are from South East and Central Asia.

Because of its colonial past Britain has had a slightly different history (Gilroy 1987; Paul 1997; Reeves 1983). After WWII, and with the beginnings of decolonisation, Britain allowed people from the colonies and ex-colonies – broadly speaking the Commonwealth – to reside and work in Britain, without visas; this was reaffirmed in the British Nationality Act (1948). A rapidly increasing influx from these areas soon provoked concern in the Labour government about 'the immigration into this country of coloured people' (cited in Carter, Harris and Joshi 2000: 24). Freedom of movement was soon restricted by the Conservatives with the Commonwealth Immigrants Act (1962), which put conditions on entry. Despite the Labour politician Hugh Gaitskell decrying the law as 'cruel and brutal anti-colour legislation', it was tightened in a 1968 amendment, which required migrants to have 'a substantial connection' to Britain consisting of a link by ancestry to a British national. In the same year the Conservative politician Enoch Powell gave his notorious 'Rivers of Blood' speech in which he warned of dire consequences of 'the future growth of the immigrant descended population', which would threaten the position of 'the native-born worker' (Powell 1969). Powell was sacked, but garnered a good deal of public support. The Immigration Act (1971) went further still, effectively restricting entry to people with parents or grandparents born in Britain. The British Nationality Act (1981) strengthened the criterion of descent, by stipulating that even those born in Britain had to have one parent who was a British citizen or permanent resident in order to be a British national.

In this trajectory of steadily increasing restrictions on non-white immigration, it is notable that (a) the legislation and political discourse on immigration rarely explicitly mentioned race or colour, although such vocabulary did get an airing in the press (Reeves 1983); and (b) as in Australia, even during the decades of maximum immigration from the New Commonwealth (which included India, Pakistan and Jamaica), white immigrants – mainly from Ireland, but also from North America, Europe and Australia – numerically exceeded or were not much less than non-white immigrants (Dumont, Spielvogel and Widmaier 2010); this remains the case today, even though in 2011 India was the single largest source of non-British-born people in the country, a position occupied by Ireland for many previous decades (Office for National Statistics 2013). It is clear that, whatever the terms being used in law and public policy, the concern of politicians and much of the wider public has been with 'coloured' immigration and the threat it has been held to pose to British national identity.

Immigration policies have sometimes effectively targeted racialised categories of legal immigrants. Often, however, the focus has shifted towards illegality, and the figure of the 'illegal migrant' – whether a failed asylum-seeker, a visa overstayer or an illegal entrant – becomes racialised, while the official line is simply that it is an issue of 'security', presented as a neutral matter. In Britain the emphasis has been and still is on restricting legal modes of immigration, but a concern with illegals is becoming more apparent; although some of these are European, many are Asian and African. In France a similar trajectory is evident as an open immigration policy between 1945 and the late 1970s was followed by a series of laws introducing restrictions and giving police more powers to check up on illegal migrants (Chapman and Frader 2004; Lentin 2004; Schain 2012).

In the United States, as non-European migration became a majority inflow, the emphasis switched to policing illegal immigration, seen as the growing problem, with some 12 million illegal immigrants currently in the country (nearly 4% of the population), of whom over 80% are from Latin America and the Caribbean, with 60% from Mexico alone (Passel and Cohn 2009). From the late twentieth century periodic cycles of crackdowns on 'illegal aliens', not explicitly racial, have targeted Mexican migrants, despite their key role in the manual labour force. The infamous California Proposition 187 (1994) proposed to prohibit illegal aliens – mainly Mexicans – from accessing state social services, although it was ruled unconstitutional by a federal court. In the 1980s and 1990s the US authorities acted to restrict the number of Haitians trying to enter the United States, fleeing political persecution; many Haitians were forcibly repatriated (Johnson 1998). The illegal alien has thus become a racialised category.

In Russia a vocal anti-immigrationist lobby has emerged since the collapse of the Soviet Union after 1989, some elements of which draw on neo-Nazi ideas and symbols (Zakharov 2013: ch. 7). The movement racialises the issue of immigration, opposing 'whites' or 'Slavic' peoples to 'blacks' or 'non-Slavic' people from the Caucasus region and Central Asia, even though about half of Russia's 11 million immigrant population in 2002 was born in countries such as Ukraine and Moldova, whose people would not generally be classified as 'black' by Russians (Migration Policy Centre 2013). 'Black' was used as a label for non-Slavs in the Soviet era and – rather than being associated with the African diaspora – has come to connote 'black market' and 'black economy' (Reeves 2013). It has propensity: cultural fundamentalism assumes that xenophobia is 'only natural' (Lentin 2004: ch. 2; Stolcke 1995).

The problem with these ideas about cultural racism and cultural fundamentalism is that they very rarely explore exactly what is at work in terms of the operation of naturalisation, biologisation and essentialisation. For example, I think the argument about cultural fundamentalism underplays the way naturalisation works. To see xenophobia as natural implies seeing each people's culture as deeply engrained and not easily shifted; it implies a strong component of essentialism and naturalisation. As Stolcke herself recognises, it also taps into an 'organicist' conception of national belonging, which assumes that the nation is an organic entity to which people belong in part by virtue of rootedness, ancestry, long-term upbringing – in short, being born and bred – and which has always been in tension with more 'voluntarist' conceptions of the nation, which link belonging to choice and willingness to participate (Stolcke 1995). But the ways in which cultural fundamentalism might actually entail the naturalisation of particular cultures, as well as the naturalisation of xenophobia as a trait of 'human nature', are left unexamined. Likewise, theories of cultural racism do not delve into what happens when culture is biologised, naturalised or simply essentialised. This is a complex area, as we can see if we look at essentialist discourses and the variety of implications they contain, some of which are more naturalising than others.

CULTURAL ESSENTIALISM: TWO EXAMPLES

1. Négritude. This was a literary, philosophical and political movement founded in Paris in the late 1920s by three visiting students from the French colonies: Aimé Césaire from Martinique, Léon Damas from Guiana and Léopold Senghor from Senegal. It was explicitly anti-colonial and anti-racist, and sought to revalorise blackness through the use of poetry and critical and philosophical writings. A key moment in its trajectory as a movement was when the French philosopher Jean-Paul Sartre wrote 'Black Orpheus', the preface to an anthology of black poetry published by Senghor in 1948. Négritude became a powerful rallying point for anti-colonial and black identity movements in the Atlantic world. It had multiple facets, not easily reduced to a simple formula, but it drew criticism for 'essentialising' black/African identity. For example, some of the writings of Senghor and Césaire seemed to characterise black Africans as essentially emotional, sensual and rhythmic: 'Emotion is Negro, as reason is Hellenic', wrote Senghor in a 1939 essay (Senghor 2003: 288).

   There are certainly essentialist tendencies here, although Senghor was more complex than this too, but the character of them is ambiguous (Diagne 2014). Occasionally, Senghor ventured into explicitly biological discourse, as when he wrote that 'the physio-psychology of the African explains his metaphysics and consequently his social life' (cited in Trimier 2003: 188). But generally he talked in terms of l'âme nègre (the black or Negro soul), seen as an essential characteristic of all African and African-descent people, but which remained vague in terms of how biological or natural it was, if at all. Césaire was more cautious, and distanced himself from Senghor's 'metaphysical' essentialism, yet in his poetry he used metaphors of roots, blood and trees to talk about Caribbean culture. At the same time, he emphasised the history of suffering that united African diaspora people and explicitly rejected biologically and geographically determinist explanations for African cultures (Césaire 1955: 20–3; Licops 2002).

2. Culture of poverty. A more complex case is the idea of a culture of poverty. This was originally introduced by the anthropologist Oscar Lewis to describe how certain 'subcultures' among poor Mexicans were characterised by a self-perpetuating set of cultural attitudes – feelings of fatalism, helplessness, lack of a sense of history or social location – that were passed on from one generation to another, locking the members of the subculture into a cycle of poverty (Lewis 1959). The idea gained some traction in conservative circles in the United States in the 1960s and, despite heavy criticism, has never gone away as a politically conservative, 'blame-the-victim' explanation for why many poor people stay poor.

By locating the causes of poverty in the characteristics of the poor themselves, critics argue that the concept ignores the bigger societal causes of poverty, while making it appear an unchangeable and essential part of poor people's lives (see Lamont, Small and Harding 2010). Although Lewis did not apply his ideas to a racial or ethnic category, they have been used to talk about black poverty in the United States by describing black family structures as 'pathological' arrangements that reproduce poverty (Moynihan 1965; Stack 1974). The concept of a culture of poverty easily becomes essentialist, but it sticks more clearly to the side of culture than does Négritude: metaphors of biology, blood and nature are not usually present, even if the culture is perpetuated across the generations.

These examples show some of the variety in cultural essentialism, and also indicate the need to explore in more detail how naturalisations may continue to operate. In what follows I therefore explore the ways nature and biology appear in racialised thinking in an everyday world where ideas of difference are apparently dominated by a language of culture. A vital clue for this exploration is contained in Stuart Hall's comment that, in current concepts of ethnicity, which are apparently only about culture, the 'articulation of difference with Nature (biology and the genetic) is present, but displaced *through kinship and intermarriage*' (2000: 223, emphasis in original). While culture may be the key mode of discourse about difference in a post-war world in which anti-racism is a key value and one of the only contexts in which 'race' easily appears, we find that concerns with biological aspects of racialised thinking remain evident when it comes to families, reproduction, children and heredity.

Race–kinship congruity

A key conceptual principle here is what I have called 'race–kinship congruity' (Wade 2012). This operates mainly in kinship systems that are built on the assumption that both parents make a contribution of physical substance to the child they conceive together; thus the way a child looks physically can be explained mostly, if not entirely, by reference to the way the parents look – or if not them, then other relatives connected by lines of descent. In English, phrases such as 'a chip off the old block' and 'like father, like son' reflect this idea (the latter phrase in a way that privileges paternal descent lines). In Latin America they say '*hijo de tigre sale pintado*' (the son of a tiger comes out striped). These phrases can refer to temperament and character as well as appearance, but they all invoke genealogical connection.

Another study of the same city reinforced this image of the importance of a naturalising discourse, but one that was intimately interwoven with a discourse about culture. Looking at white mothers of mixed-race children, France Winddance Twine found that they were often assumed by local black people not to have full 'empathy' with their children: they had never experienced racism themselves, and could only 'sympathise' with their kids. Black mothers, on the other hand, were assumed to 'possess maternal competence by virtue of belonging to the same race as their children' (Twine 2000); even if the black mothers had lighter-skinned, mixed-race children, that racial bond was assumed to be there. Now, biology, as such, does not explicitly enter the picture – empathy could be a result of sharing certain experiences of racism in Britain. But the automatic assumption of maternal competence built around shared blackness indicates a concern with consanguineal kinship connections. Some white mothers feared that their own maternal bond with their children could be disrupted by racial difference – culture and kinship were working against each other and produced only sympathy. For black mothers the kinship connection was working in the same direction as the cultural one, and they strengthened each other to produce empathy. Biology – understood as kinship links of 'blood' – was not operating alone here and was not determining much at all, but it was clearly part of the way belonging and relationships were being worked out.

Assisted reproductive technologies (ARTs)

ARTs involve the techniques of the donation of ova and sperm, in vitro fertilisation (IVF) and gestational surrogacy. The question of natural and cultural kinship connections is made particularly explicit by these technologies, as in certain combinations they can separate out the genetic parents (the donor of ovum or sperm), the birth parent (the woman who gestates the embryo) and the social parents (the people who bring the child up as its socially recognised mother(s) and/or father(s)). The issue of race–kinship congruity also takes on specific forms, because, although an ART child's physical appearance in relation to his/her social parents will always be a matter of concern, when racialised appearance is involved the concern is often sharpened.

In general, the use of ARTs tends to be the object of state regulation to ensure ethical practice in relation to the use of human gametes, although the rules vary quite widely from one country to another (Edwards and Salazar 2009). A common assumption in ART practice is that race–kinship congruity should, by default, be the norm. For example, it is assumed that a black couple trying for an ART baby will want gametes donated by black people (according to local definitions of black).

In Britain, racial matching in ART clinics was at one time official policy, obeying the rationale of mentioning race as part of an inclusive anti-racist strategy. Interestingly, over a few years the policy dropped any mention of race, opting instead for colour-blindness. In 2002 the Human Fertilisation and Embryology Authority issued a guideline saying that clinics should 'strive as far as possible to match the physical characteristics and ethnic background of the donor [of gametes] to those of the infertile partner'; they added: 'those seeking treatment should not be treated with gametes provided by a donor of a different racial origin unless there are compelling reasons for doing so' (HFEA 2002). In 2003, in the sixth edition of their Code of Practice, the word 'racial' was dropped, leaving 'physical characteristics and ethnic background'. From 2007 the Code dropped all reference to ethnicity and race, simply saying that the clinic should consider factors which might cause 'significant harm or neglect' to either the child to be born or other children in the family. Race is only mentioned to say that no discrimination should be made on that basis (HFEA 2014: section 8). Nevertheless, donors are advised that data will be collected which may ask for their 'ethnic group' as well as their 'physical characteristics' (HFEA 2012).

In Spain the law dictates that gamete donors 'should have the maximum phenotypic and immunological similarities and the maximum possibilities of compatibility with the receiver and her family environment' (Jefatura del Estado 1988: Article 6); to facilitate this, further legislation defines the data that are to be collected on donors, which include skin colour, eye colour, hair colour, hair texture, blood group and 'race' (Ministerio de Sanidad y Consumo 1996: Annex).

What drives this kind of legislation? To a large extent, it obeys the aims of anti-racism, which we examined above under the rubric of 'enumerating race for inclusion'. Racial matching in ART practice follows social policy guidelines on adoption that have been influential – but also controversial – in the United States and Britain, according to which it is in the 'best interests of the child' to be brought up, if possible, in a household that is racially matched to the child. The idea, contested by some, is that ensuring that non-white children have a non-white family environment will be a supportive tactic when racism is a possible threat in the wider society, while placing a non-white child in a white family could rob him/her of needed support and even cause conflicts of identity (Fogg-Davis 2002; Hollingsworth 1998). In both Britain and the United States the law provides that adoption agencies cannot deny or overly delay an adoption for reasons of racial–ethnic matching, but practice in such agencies is generally to match where possible. The controversial character of racial matching in adoption policy is probably behind the HFEA's dropping of the explicit mention of race or ethnicity and the implication that the institution should not be regulating such matters – even if donors' ethnic and physical data are still recorded.

These policies fall within an anti-racist agenda, but they are different from the simple enumeration of race, in order to monitor racial inequality and seek to address it. They bring into play race–kinship congruity and the hereditary aspects of racialised appearance because the policies in effect ensure that ART children *look as if* they belong to the family in which they live, including – indeed, principally – in terms of racialised appearance.

In practice, people who use ARTs adopt different attitudes to racial and ethnic matching. Often the default assumption is that matching is the norm, maintaining race–kinship congruity in a fairly simple way. This is evident in cases where ARTs 'go wrong', in accidents of racial mismatching. One 2002 case in Britain resulted in a white woman giving birth to 'black' twins, when a black man's sperm was mistakenly used to fertilise her eggs. Much of the subsequent press coverage reported the incident in terms of 'shock' and 'horror' and the traumatic effect on the mother of giving birth to 'black' children

Race is not mentioned, and this is in keeping with standard practice in Norwegian society, where the word is rarely used and, for example, non-white immigrants are talked about in terms of cultural difference (although the possible existence of racism is acknowledged). Likewise, what matters for the adopted children is culture. Culture works a bit differently for non-white adoptees and non-white immigrants. The former are radically transformed by cultural kinning; the latter are seen to retain their culture of origin, and even transmit this to their children, born and bred in Norway. In the first case, the difference in appearance is construed as signifying nothing; in the second, it is construed as signifying engrained cultural difference.

However, the difference in appearance of the adoptees seems not to be completely erased: somehow it remains as an unavoidable reminder, especially outside the immediate family context, when people may comment, uninvited, on the child's appearance. Difference reveals itself in the concern some parents show for the 'original culture' of their adopted children. These were generally adopted as small babies, raising the question of what 'original culture' – or 'birth culture' as it is sometimes called in the United States (Volkman 2005a: 97) – means in this context. But in practice, some families participate in the social get-togethers of organisations such as the India Association or the Colombia Association, where people dress up in the clothes and eat the foods associated with these countries of origin. Some families also go on 'motherland' or 'roots' tours to visit the countries of origin of their children, with the idea that this will be of interest to the children and support them in some way in what is, in effect, being conceived as a 'mixed-race' identity, although such terms are never used. The children themselves rarely engage strongly with the essentialist idea of an original culture with which they should identify: they feel Norwegian. Neither do they engage strongly with adoptees' associations, in contrast to some other contexts in which adoptees from specific countries, such as Korea, have formed international organisations, often facilitated by the internet (Volkman 2005b).

It seems clear, however, that the concern with an original culture is a response – disavowed of course – to the racial difference that is made evident on the bodies of the adopted children. Sometimes that overt racial difference is seen to manifest itself in other ways. In Spain, for example, some adoptive parents comment on the 'natural' way their adoptive children dance with an 'African rhythm' or eat their food delicately with their hands, or act energetically (Marre 2007). As we saw in the case of ARTs, the concern with biology is actualised in a discourse of culture; biology does not operate alone. But it is an illustration of the way questions of racialised appearance, inherited biologically, surface in a world where the dominant discourse of difference is about culture.

Latin American examples

If we are interested in the way ideas about biology, nature, heredity, blood and bodies appear in a racial discourse that is said to be dominated by culture, then Latin America is an interesting place to look. While race is always a social construct, in this region it is generally agreed to be a particularly culturalised construct. As we saw in Chapter 1, in Mexico *la raza* is used to refer to Mexicans in general, in all their variety, united by both a common history and a common mixture. In many areas of the region the difference between an *indio* (an 'Indian' or indigenous person) and a mestizo is a difference established by place of residence (rural versus urban), clothing, language, education, food habits and occupation rather than one based on racialised appearance or ancestry (Weismantel 2001): many mestizos in the Andes and Mexico have substantial indigenous ancestry and look very like people classed as indigenous.

In Argentina the term *los negros* is used by middle-class people to refer to lower-class people living in the *villas* (the low-income settlements or 'slums') who are seen as 'dark', even though to many European eyes they would look rather white. In many countries a simple measure of skin tone cannot distinguish between the racial categories people choose to identify themselves: different people with the same medium brown skin identify as white, mestizo, indigenous and black (Telles and Flores 2013). Racial and colour identifications are also contextual, meaning that people will use different terms to describe themselves and others depending on to whom they are talking and in what circumstances. In Colombia or Brazil, the same individual dressed shabbily is likely to attract a darker colour term than when looking wealthy, and economic success often leads people to opt for lighter colour terms for themselves and their children: 'money whitens', as the saying goes (Schwartzman 2007).

Race always has flexible aspects as a mode of identification, being mediated by aspects of class and culture, and constituted through performance, but in Latin America these aspects of race are more marked than in, say, Britain or the United States, where criteria of racialised appearance and ancestry define some rather clearer categories. But this does not mean that appearance and ancestry are irrelevant. For the elite, especially, questions of genealogy can assume great significance. A very Northern European-looking person is liable to be classed as *blanco* (or *branco* in Portuguese) under most circumstances; and a very African-looking person will likely be classified as *negro* (or another similar term such as *preto* or *moreno*). Despite the very cultural character of race, where bodies and biology seem to play a relatively muted role, there is plenty of evidence that, as usual when race is at issue, both nature and culture work together. Here are some short examples of this.

1. **The racialised body.** Whatever the indeterminacies of colour and identification in Latin America, it is hard to deny that colour and other aspects of racialised appearance (such as hair type and facial features) remain extremely important in people's lives. Phenotypical features do not combine into a system of reliable criteria for assigning or claiming a particular category label, but they certainly carry a powerful social and emotional charge. In Brazil, for example, plastic surgeons regularly offer to correct the 'Negroid nose'. This facial feature is not linked only to 'blacks' as a category of people: many people who think they have the trait would not see themselves or be seen as black. But it is clearly linked to *blackness* and its correction is linked to the widespread aesthetic preference for a more European appearance. On the other hand, women also regularly request to have a Brazilian *bunda* (bottom), which involves enlarging and rounding the buttocks to achieve a look that is partly associated with an African female body, but that has become a national icon (Edmonds 2007, 2010). The value attributed to blackness varies – although when

it is positive it is phrased as 'Brazilian' rather than 'Negroid' – but the concern with the racialised body is undoubted.

In Mexico skin colour is very indeterminate in people's identifications, but a good deal of weight is given to where people sit on the spectrum between *güero* (blond) and *moreno* (brown). As in much of Latin America, people are interested in 'how the baby came out', that is, how dark or light it was at birth (Sue 2013). One Mexican woman recounted that her parents had no photos of her from when she was born, because she came out rather dark. She said her parents had told her: 'So we were waiting, because also you were born a little bit ugly, and black, so we waited for you to grow up a little bit till you got better and changed.' Another woman recounted that her aunts 'thought my father was really dark [and] that it was terrible to get married to a dark-coloured person instead of improving one's class' (Moreno Figueroa 2010: 394, 396). Other women recounted that they felt their racialised appearance was a factor in their emotional relationships in the family, creating feelings of having been 'slighted', because they were less pretty than their more *güero* siblings (Moreno Figueroa 2008).

Colour is an uncertain quality: three Mexican women discussing the colour of the son of one of the three cannot agree whether he is *moreno* or quite what shade of *moreno* he is ('light *moreno*'?); yet colour is important one way or another, and the question they are pondering is why the father persistently wonders why his son came out *moreno*. The mother says: 'And I say to him: "Well, because he's your son"' – in her eyes, her husband is also *moreno*. In other words, like father, like son. The women think the father is indulging in a common fantasy, in which the baby will come out nice and *güero*, as if by magic (Moreno Figueroa 2012: 172). These questions of racialised appearance, then, are always lodged in questions of resemblance to family, which are necessarily questions about inheritance and naturalised connections between generations.

2. '*Preto* **is a color;** *negro* **is a race'.** In Brazil the official categories for 'colour or race' are *branco* (white), *pardo* (brown), *preto* (black), *amarelo* (yellow, meant to capture Asian Brazilians) and, since 1991, *indígena*. This category system is not the only one used. It exists alongside more everyday usages, which, although they can produce dozens of colour terms, in practice tend to centre on *branco* and *moreno*, with *pardo*, *preto*, *negro* and *moreno claro* (light brown) being used with less frequency. A third system divides *branco* from *negro* (understood as including the official categories of *preto* and *pardo*), a binary classification driven by both the black social movement and the state anti-racist agenda, which agree on the tactic of comparing all non-whites to whites in measuring racial inequality (Telles 2004: 81–7).

The saying '*preto* is a color; *negro* is a race' (Santos et al. 2009: 794) is an everyday usage, which hovers between the census categories and the political categories, but also does something a little different, as the following exchange between an anthropologist and a black (*pardo*) woman from Rio shows:

ANTHROPOLOGIST: So, saying *negro* and saying *preto* is the same?
INTERVIEWEE: It's different because *preto* is the colour [she rubs the skin of her arm] while
  negro is the race [she passes her hand close to her face as if to refer to facial features].
ANTHROPOLOGIST: But what is race?
INTERVIEWEE: Race is race. It's your … origin [pointing at her face again].

This woman, certainly influenced by the black political movement usage of *negro* to encompass all non-whites, is not referring only to this. She is making a distinction between people who have brown skin, which may have come from a number of different ancestral sources, and people who have brown skin *and* facial features that show they are of African descent and thus might legitimately identify with the political category *negro* (Sheriff 2003). This is a distinction that splits the *pardo* category into brown-skinned people who don't look Afrodescendant and brown-skinned people who do; the latter are included with the *pretos* to form a category of *negros*. This is a distinction that rests on a concept of descent and the inheritance of a particular racialised phenotype.

A long time ago Brazil was said to have racial prejudice of 'mark' (appearance in the broadest sense) in contrast to the race prejudice of 'origin' of the United States (Nogueira 2008 [1959]), a characterisation that has stuck in many respects, as it seems to capture the US 'one-drop rule', which assigns a black identity to anyone deemed to have 'one drop of black blood' versus the Brazilian emphasis on skin colour. However, in this Brazilian saying we see a different emphasis: there are ideas about both mark and origin operating together. Again, racialised appearance entrains ideas about the biology of heredity.

3. **Like mother, like son.** In Guatemala there is a powerful distinction between *ladino* and *indio*. The former is the local term for non-indigenous and encompasses whites and what in other countries would be called mestizos. The distinction is, for historical reasons, a good deal sharper than the distinction between *indio* and mestizo in neighbouring Mexico. It is also a divide that has been characterised by severe conflict from the 1960s through the 1990s, as an authoritarian government harshly repressed and attacked indigenous Maya communities seen as subversive collaborators of left-wing guerrilla movements.

Diane Nelson (1999: 231) recounts the case of a young *ladino* man who told her how his mother had always instructed him never to sit next to an indigenous woman on the bus. She was afraid that other people might assume he was the indigenous woman's son. To Nelson the man looked non-indigenous, but clearly his mother saw something in his appearance that, when placed in the context of a possible genealogical relationship, suggested by the physical proximity of sitting side by side, would say '*indio*' to other observers. His appearance alone was not enough: he had to be next to an indigenous woman of the appropriate age for the kinship connection to be suggested; ideas about biology were only activated in the context of the right cultural cues. In Guatemala immediate cues of proximity on the bus are supplemented by the fact that many *ladinos* say that they can always tell if someone is indigenous by certain physical markers, but the way *ladinos* behave 'suggests the terror of not being able to tell' (Nelson 1999: 231): that is, there is a pervasive concern with the boundary between *indio* and *ladino* and the possibilities of blurring it.

4. **Inheritance and the body in Colombia.** In Colombia a certain style of music associated with the Caribbean costal region of the country became nationally (and internationally) popular from the 1940s onwards – styles such as *cumbia* or, more generically, *música tropical*. The Caribbean coastal area is associated strongly in the national imaginary with tropicality and blackness, compared to the Andean interior, which is associated with highland climate, lighter-skinned mestizos and whiteness. (Indigenousness is associated with both lowlands and highlands, but seen as located in peripheral areas in both cases.)

For a country in which elites had generally marginalised lowland blackness as a component of national identity, preferring to emphasise highland whiteness and mixture, it was striking that musical styles associated with tropical blackness should achieve the status of a national icon and be played by musicians from all over the country (Wade 2000).

One musician, a light-skinned man from the highland city of Medellín – reputed for its relative whiteness – in explaining his predilection for this tropical music, said he liked it because he had a *corazón de negro* (a black man's heart). This suggests a version of *mestizaje* in which people are not simply homogeneous fusions of different ancestries and traditions, a kind of soup created according to the blending theories of inheritance that were popular from the Greeks to Darwin. Instead, people are combinations of durable components, a mosaic created by the particulate theory of inheritance that came with Mendel. Racialised ancestry was reflected in specific tastes and abilities (Wade 2005). Other mestizo and white Colombians in Bogotá sometimes also linked racialised inheritance and their personal characteristics: one man who had spent some time living in the Caribbean coastal region said, 'I was much more akin than what I initially thought to their food, their housing, the sea. So I said: this must be because of a black ancestor.' Others said: 'When I hear the drums I feel like I am from over there: I guess I must have something black in me [*algo de negro debo tener*]' (Schwartz-Marín n.d.).

Ideas about the role of inheritance varied in Colombia. One young Afro-Colombian rap artist, part of a small black activist organisation in the city of Cali, rejected biologically reductionist explanations which proposed that black people 'have a special body or X type blood'. Instead, he said 'the drum, the sound, the music are indivisible from the African man' and 'if a person has, for 2000 years, been listening to the drum … something has to happen in the body, in his mentality, in his rhythmic way of being, of moving'. But an environmentalist explanation like this is still a powerfully naturalising one, which engrains rhythm into the black African body and hence into his own Afro-Colombian body. Other Afro-Colombian artists adopted a more biologising discourse. For one dancer, the white middle-class students he had grown up with were 'lifeless' and 'like sticks'; this was 'in the blood, in the descent'. In contrast for him: 'when I hear the boom-boom-boom of the drum, I run out to have a look … it calls me as if by telepathy'. Even for this man, biology was not the only factor: he admitted uncertainty about whether musical and dance abilities were in 'the blood or the tradition'. But for both men naturalising discourses about inheritance and descent were certainly present in thinking about racialised difference (Wade 2002b: 101).

CONCLUSION

The purpose of this and the preceding chapters has been to interrogate the related ideas of (a) a shift from 'biology' to 'culture' in thinking about race; and (b) a shift from 'race' to 'culture', in which race itself disappears, to be replaced entirely by a discourse of culture or to be submerged and masked by this discourse, while remaining 'buried alive'. In Chapter 4 we saw that thinking about race as a biological reality did not disappear so readily and has returned in some respects, although not always named as such and going under titles such as 'bio-geographical population'. In this chapter we have seen that race may remain as an institutional presence (in the practices of the state, the medical world, etc.) often as part of an anti-racist agenda that recognises race as a social category, and enumerates it as part of an inclusive mission to eradicate or at least ameliorate racial inequality.

When it comes to exclusion, institutional practice may effectively discriminate against racialised categories of people – categories that we as analysts can identify as racialised – but such practice is very rarely admitted to have racial dimensions, precisely because doing so would contradict the anti-racist mission of inclusion, which is the legacy of post-WWII global agendas. Occasionally we can see the slippage, as when an official report concluded in 1998 that the Metropolitan Police in London was 'institutionally racist' – a conclusion that was tempered by saying that the racism was unintended (despite admitting that it was part of police 'canteen culture' to stereotype black people as thieves) and an allegation that was rejected by the then Commissioner of the Metropolitan Police, Paul Condon.

Making race publicly visible, then, is often about showing the cracks in the anti-racist mission and demonstrating that policy measures that are ostensibly about 'security', 'fighting crime' and 'controlling illegal immigration' may have racially discriminatory effects. But showing how racism and racial thinking enters into both institutional and everyday practice requires more than showing that certain practices have racially discriminatory effects, which can always be excused as 'unintended'. We need to show how racial discrimination and racial disadvantage are reproduced through multiple practices, with complex and often opaque motivations at work, some more racially inflected than others but with racism rarely an explicit motive – after all, who today admits to being a racist?

This question of the way racial difference and disadvantage are reproduced is a topic we will address in the next chapters in relation to specific regional contexts. But in the latter half of this chapter we began to see how ideas about race enter into everyday practices and motivations, frequently via the family and kinship. Here we saw, through concrete examples, how ideas about inheritance, biology, bodies and blood remain significant aspects of racial thinking in an era of cultural discourse about difference. This helps us to identify these phenomena as racial, as opposed to just cultural ones. As Wieviorka (1995: 142) has said, if people refer only to 'nation, religion, traditions and, more generally, culture, with no references to nature, biology, genetic heritage or blood, it is preferable not to speak of racism'.

FURTHER RESEARCH

1. Explore the uses of racial enumeration in censuses: start with Morning (2008), then look at Loveman (2009) and perhaps also Nobles (2000); see also Owen (2001). What drives the use of race in censuses? How is it done?

2. There is a lot of debate about the pros and cons of using racial and ethnic descriptors in medical practice. Starting with the sources in the relevant text box in this chapter, delve into these debates and draw your own conclusions.

3. Choose a specific country and research its changing immigration and citizenship policies. The government websites of individual countries may be a good starting point. The Migration Policy Institute website is a useful resource (http://www.migrationpolicy.org/). The Multiculturalism Policy Index (http://www.queensu.ca/mcp/) has data on this and other issues, and it has links to more information about citizenship. The OECD's website on migration has a lot of data and analysis (http://www.oecd.org/migration/).

4. Go to the main website of the Ku Klux Klan (http://www.kkk.com/). Try also the websites of Britain's National Front party and the English Defence League. Do they admit to being racist? How do they justify their beliefs?

PART II

RACE IN PRACTICE

6 Latin America: mixture and racism

6.1 Introduction

In Part II the approach shifts away from the broadly chronological towards the ethno-graphic. In the previous chapters we traced the changing meanings of race, with an eye to the constant interweaving of ideas of nature and culture, and with the aim of nuancing conventional chronologies that – to oversimplify – traced a broad historical sweep from race as culture, to race as biology, and back to race as culture, with race itself disappearing in the last phase. From this point in the book the focus is on the way racial inequality and racial difference are reproduced in changing ways in everyday practices in specific places. The idea is not to give regional overviews, but to focus on specific ethnographic cases in their regional context. Three preliminary points need emphasis.

Racial categories as relational

First, racial difference is not just about disadvantaged racialised groups, but about a set of relational racialised identities: white, mixed, black, indigenous, Asian, etc. are all equally significant; these categories or identities gain their meanings and power in relation to each other. This emphasis derives in part from recent research focusing on whiteness, as a category or social position that, in the past, has not always received much attention, as the study of race has typically been a matter of white researchers investigating groups seen as 'other' and usually disadvantaged or oppressed, a viewpoint that tended to take whiteness for granted or assume its privileged status. Greater attention to whiteness has shown that it is also a histori-cally changing and unstable category – one that, in the United States, for example, during the late nineteenth and twentieth centuries absorbed categories previously defined as non-white, such as southern Europeans, the Irish and Jews (Brodkin 1998; Kolchin 2002; Roediger 2006) – and a heterogeneous one for which privilege is not necessarily automatic (Hartigan 1999).

Racial practices as relational

Second, understanding racial difference and inequality requires seeing race as relational in a different sense and embedding it in a local cultural context, paying attention to class, politics, space, gender, values and morality. It is no good assuming that race can be understood as a set of stand-alone ideologies, prejudices, images and classifications, which operate independently to shape life chances, opportunities and freedoms. Race is always constituted in relation to and interwoven with other processes and ideas; it forms an integral part of other practices, such as bringing up a family, working, having fun, finding a place to live, migrating, doing 'the right thing', being 'a good person' or just 'a nice person' and so on. This is because race is 'a social construction', but this does not mean race must therefore be 'constructed' by other more 'real' factors, such as class. As I argued in Chapter 4, all human social life is, by definition, socially constructed and even things that in social theory have often been given the status of an independent, objective, determining force, such as class, must be conceptualised as emerging through the social organisation of things, people and ideas. And it is precisely because all these factors are socially constructed that they interpenetrate and shape each other. Race is relational: it is part of a complex assemblage of practices, objects and ideas, and it draws meaning and power from its relation to other elements, as well as conferring meanings and power on those elements (M'charek 2013).

RACE AND CLASS: WHICH IS 'MORE IMPORTANT'?

A lot of theorisation about race has debated the relationship between race and class, often in terms of which has greater powers of determination in theoretical terms. Views range from a simple Marxist position, which holds that economic forces are the real material basis that determines the changing shape of race, understood as an ideological façade, through the more nuanced Marxist theories that accord greater 'autonomy' to race, to 'racial formation' approaches, which see race as a factor that can shape social change as powerfully as class (Back and Solomos 2000: chs. 9, 10, 13; Kolchin 2002; Miles 1989; Omi and Winant 1986). These debates have waned somewhat, but it is still often assumed that 'class and race are concepts of a different order' and thus that 'they do not occupy the same analytical space' (Fields 1982: 150), because class is a material reality, whereas race is ideology. I think it is more productive to see both race and class as involving material and ideological dimensions; or, avoiding this unhelpful dichotomy, to see that race and class both involve the ongoing arrangement or assembling of people, ideas and things into the constellations of relationships and processes that constitute social life.

For example, if we think back to the theme of immigration (see Chapter 5) it is clear that 'white immigration' and 'non-white immigration' raise different issues in people's minds, in politics and in policy in many Western countries. It is not very helpful to think about this as a basically material issue of competition for jobs and housing, in which race enters as a secondary ideological factor, because race organises and structures so much of the practice and discourse around immigration. Race is often influential in constituting what counts as 'legitimate' competition in the first place, because non-whites may be seen by white residents as people who are not as deserving as whites of a place in the country. Race is often influential in defining who is likely to count as 'illegal', because in certain contexts illegals are more likely to originate from non-white areas and classes, such that, in the country of destination, race becomes a marker of illegality in the eyes of the authorities.

An example of the way race interweaves with other practices comes from the common trope among white people in British urban areas with large Asian immigrant populations that 'their food stinks' or 'smells funny'; the same comments can be found in Australia

(Cerwonka 2004: 199) and the United States. One US internet user in 2011 asked Yahoo Answers, 'Why do indian people always smell like curry, without fail?', adding, 'I hate to be racist, but is it cause they use curry for washing powder? I just dont get it!' The various answers on the forum mostly responded in terms of Asian food and cooking practices, one adding that whites smell of sour milk to Asian people and another wondering if Americans smell of hotdogs and mayo!

The point here is that race is embedded in everyday ideas and practices around food, and values about what is nice and nasty, and this is part of what organises patterns of residential and commercial segregation, not easily separable from, say, the price of housing (Slocum and Saldanha 2013). The preference of some white British people not to live in Asian areas because of the food smells is a choice interwoven with others made on the basis of house prices; such choices help constitute neighbourhoods with concentrations of Asian immigrants, a pattern which itself may drive down house prices in the area. Ideas about food also shape practices of assimilation: in Britain some of the middle-class descendants of Asian immigrants in Manchester restrict the consumption of Asian food to special occasions, on the basis that the smell is too strong. One respondent in the Yahoo Answers forum said, 'I am from India and I myself choke on Indian guys' odor in elevators … I don't want to smell bad either, so I rarely cook Indian and have switched to western diet.'

Places as connected

The chapters in Part II are about specific places, because I think it is necessary to grasp local contexts and their histories in order to understand how race operates to differentiate, include and exclude. In this book regional coverage has to be selective, so I focus on regions where race is clearly an important factor. It is important, however, to see that these contexts (and others) are interconnected and have been for a long time. Of course, migratory currents have created diasporas centuries old, but more specifically, ideas about race have circulated long and widely, especially in the Atlantic world, and not just in terms of the international science of race of the late nineteenth and early twentieth centuries.

Black Brazilians travelled to West Africa and England in the late nineteenth century and, on their return to Brazil, influenced the character of Afro-Brazilian religious practices (Matory 1999). British colonial rule in Ghana generated a racialised consciousness there that laid the basis for Ghanaian independence leader Kwame Nkrumah's participation in the cosmopolitan internationalist pan-Africanism that had been developed earlier in the century by black intellectuals in the Caribbean, Europe and the United States; this in turn laid the basis for Ghana's emerging role as a black heritage tourist site, commemorating slavery and pan-Africanism, with African Americans common among the visitors (Pierre 2013). Black musicians and artists travelled widely, circulating between the Americas, Europe and Africa (Gilroy 1993; Yelvington 2006). In short, blackness is a transnational construct, formed in ongoing international dialogues (Clarke and Thomas 2006). By extension, the same is true for whiteness and other such racialised identities, which have roots in the inherently transnational processes of colonialism (Moreton-Robinson, Casey and

Nicoll 2008). The fact that one finds similar comments about the smell of Asian food in Britain, Australia and the United States is a small indicator of this transnational process.

6.2 Latin America and *mestizaje*

In this chapter we look at some ethnographic examples from Colombia and Brazil for insight into how racial difference and racial inequality are reproduced in a context in which *mestizaje* ('mixture' of origins, peoples and cultures) is a deeply rooted historical process and a dominant trope for national identity. When mixture creates societies in which most people do not identify as either black, white or indigenous – and in which many people do not see racial identifications as that significant for them – what happens to racial difference and inequality?

The nature of the colonial regimes in Latin America led, in many areas, to mixture between Europeans, Africans and indigenous people, and to the social recognition of intermediate categories of mestizos, who also mixed extensively among themselves (see Chapter 2). These patterns were regionally uneven: mixture was more extensive in places such as Brazil, Colombia, Venezuela and Mexico, where – in rough terms, given the vagaries of defining racial identities in Latin America (see Chapter 5) – white and mixed people together form the vast majority, with mixed people varying between 40% and 80% of the total, while black and indigenous populations together constitute between 5% to 15%. In contrast, in the central Andean countries (Bolivia, Ecuador and Peru) and in Guatemala indigenous peoples still constitute between a third and half of the population.

In the nineteenth century slavery was abolished without a major conflict – in contrast to the traumas of the US Civil War – even when abolition happened very late, as in Brazil (1888) and Cuba (1886). During the same period governments also tried – with varying success – to undermine the special status that colonial rule had given to indigenous communities. The region's governing elites espoused European liberal values, which meant in principle that indigenous people should become citizens, legally the same as everyone else. In practice, this meant negotiating the usual tensions between ideals of equality and realities of hierarchy (see Chapter 3), with indigenous, black and many mestizo people confined to the lower rungs of the ladder.

In the first half of the twentieth century some intellectuals and elites, concerned to build a distinctive national identity, valorised *mestizaje* as the basis for that identity, challenging Eurocentric racial theories about the degeneracy of mixture. This was evident in Colombia and most notable in Mexico, where in the 1920s José Vasconcelos elaborated the concept of the *raza cósmica* (a 'cosmic', mixed race, already present in Mexico and which presaged the global future), and in Brazil, where in the 1930s Gilberto Freyre wrote in positive terms about his country's mixture. He saw mixture as related to tolerance, and his ideas contributed to the image of Brazil as a 'racial democracy', which became part of the official representation of Brazil under the dictator Getulio Vargas.

This idea of mixture as a positive basis for national identity – which is not limited to these countries, but can also be found in Bolivia, Ecuador, Peru and Guatemala among others – by no means eradicated racial difference or hierarchy. On the contrary, *mestizaje* was seen as valuable because it was assumed that it led towards the assimilation of black and indigenous peoples into a mixed population that would, eventually and especially if helped by European immigration, tend towards the whiter end of the spectrum. Latin American elites may have wanted to avoid the idea that their national populations were destined for degeneracy, but they were also in thrall to the idea that whiteness was linked to modernity and progress (Martínez-Echazábal 1998; Rahier 2003; Stepan 1991; Wade 2010).

Critiques of 'racial democracy' and of *mestizaje*, seen as a homogenising ideology that marginalised indigenous and black identities, mounted during the twentieth century. Critiques came from academic sectors, which amassed evidence of racial inequality and discrimination, and from indigenous and black activists. The latter were inspired by long histories of resistance and struggle dating back to early colonial times, and by their own experiences of racism. They drew on post-WWII challenges to *mestizaje* as a national project sidelining black and indigenous identities, or appropriating them as symbolic supports for positive images of national *mestizaje*; and they took inspiration from decolonisation and the global emergence of black and indigenous movements (Safa 2005). These critiques contributed to varied regional trends towards the official recognition, from the late 1980s, of indigenous and black communities as part of multicultural nations, and in some cases the extension to them of special rights (e.g. around land and sometimes education). These changes have marked a turning point in the visibility of black and indigenous peoples, but have not necessarily brought corresponding material improvements and have also created some backlash from within the white and mixed majorities (Hooker 2009; Sieder 2002; Wade 2010: ch. 6).

6.3 Colombia: racial discrimination and social movements

When I started ethnographic research on race in Colombia in the 1980s there was little attention to questions of race and racism in the country, especially in relation to *los negros* (the blacks). Individuals might agree that some people were racists – who, it was joked, would countenance 'nothing black, not even the telephone' – but it was anecdotal stuff, with racism often attributed to particular intolerant individuals. Black people themselves often denied to me that they had suffered racism, as if being the target of discrimination was admitting to having shameful personal faults that merited such treatment. Race was an uncomfortable topic for many people; it seemed to be subject to 'cultural censorship' (Sheriff 2001), a matter better left unmentioned.

Small-town Colombia

It was in this context that I carried out fieldwork in the small rural town of Unguía, near the border with Panama, with a view to exploring racial identity and discrimination (Wade 1993). The town was in a frontier zone that had been colonised at different times by people from three distinct regions of the country. Divided by geography and history, Colombia is often talked of as a 'country of regions' and race is one dimension of this regionalisation. Unguía had been colonised originally in the early twentieth century by black and dark-skinned mestizo people – 'Costeños' – from the Caribbean coastal region to the east, a hot lowland area with a very heterogeneous population that historically has had a significant black and indigenous presence. Soon after came black people – 'Chocoanos' – moving north from the Chocó province of the Pacific coastal region, a low-lying, hot and very humid area, very poor and inhabited mainly by the descendants of the slaves taken there by the Spanish to mine gold; it is the 'black region' in Colombia's national imaginary. Rather later, from the 1950s, migrants from the highland province of Antioquia began to arrive in force: the 'Antioqueños' came from an economically power-ful region, based on coffee, industry and commerce, and some, but by no means all, of them had economic resources to invest in the frontier economy of Unguía, based on agri-culture and cattle raising; they were mainly light-skinned mestizos and whites and had the reputation, locally and nationally, of being rather racist, or at least very proud of their lightness and their regional success story.

What was striking was the dominance the Antioqueños had quickly achieved over the local economy, being the biggest owners of land and cattle and monopolising the major commerce of the town. There were poor Antioqueños, but among those who had arrived with little or nothing, some had made advantageous connections with their compatriots and had quickly become wealthy. The Costeños were mainly poor and mid-dling landowners and labourers, while the Chocoanos tended to be labourers, service providers and small-scale gold miners, with a small handful of farmers, partly dependent on looking after the cattle of Antioqueños. They were also low-level public employ-ees, such as nurses, teachers and municipal administrators, as appointment to these jobs often depended on connections to the provincial administration of the Chocó, which was controlled by Chocoano political bosses in Quibdó, the province's capital. The black Chocoanos were clearly the losers in the local economic race (for a similar case in Ecuador see Whitten 1986).

A great deal of this economic stratification was due to the character of the regions that each set of people came from: the Chocó was very poor, with low educational levels, subsistence-level agriculture and gold mining, and little commerce; in contrast, Antioquia was one of the richest and most powerful regions in the country, with long-standing traditions of intensive agriculture and commerce. The Chocoanos made use of flexible and extended kin networks – generated in part by a tendency for men and women to have several partners over their lifetimes and, occasionally, simultaneously – to move around between unstable economic opportunities, including subsistence farming, small-scale gold mining, fishing, wage labour and administrative jobs. In Unguía this could be seen in the high frequency of extended households, made up of various kin, and in the way the few small Chocoano corner shops were undermined by the demands of relatives to obtain goods on credit. The Antioqueños in Unguía tended to have nuclear family households, based on cross-cutting portfolios of land, cattle and commerce; they too had to sell goods on credit, but they were better able to withstand defaulters.

However, part of the reason for the Antioqueños' success lay in their tendency towards ethnic and racial exclusiveness. They preferred to cooperate with other Antioqueños, because in a risky world, common regional origins were an indicator of trustworthiness and shared values of ambition and progress. Credit, business partnering and cattle- and land-sharing deals tended to be made within the group. This was underlain by racial ideas, because the least trusted in all of this were the Chocoanos, whom the Antioqueños stereotyped as being lazy, untrustworthy, not interested in progressing, inconstant, having disorganised family lives, and being overly interested in womanising and partying. The flexible family arrangements that worked as a kind of economic resource for Chocoanos were seen as evidence of irresponsibility by the Antioqueños.

True to its location in the Colombian nation, Unguía was a locus of *mestizaje*: Costeños and Chocoanos intermarried – in legal or common-law unions – a great deal; the Antioqueños were the most exclusive in this respect, but some of them formed unions with non-Antioqueños. Because of this, regional identities were not sharply defined, and racial ones less so: although some people were clearly 'black' and others 'white', there were many intermediate people, ambiguous in terms of racial categories. Although I witnessed drunken conflicts with racial insults flying and although people easily indulged in stereotyping along regional and racial lines, there was also a lot of everyday interaction in which such identifications seemed of little relevance. The children of Antioqueño migrants, in particular, were often integrated with their Chocoano and Costeño peers, in the school, the sports teams and the dance halls, even if, over time, their economic trajectories went higher, due to the resources of their parents.

Yet the overall effect of the process of colonisation of the area was to reproduce and indeed intensify a sense in which modernisation, development and progress lay in the hands of whiter people who brought these values to the benighted frontiers of the nation, while the local *negros* were being left behind. The Antioqueños were making money as individual frontier adventurers, but they were also enhancing an image of themselves as successful entrepreneurs, an image closely related to their relative whiteness and reinforced by their rapid economic dominance over the local black residents. A few of the local blacks were successful, mainly as farmers, but also with connections to the political machine of the Chocó province. But they were heavily dependent on cattle-sharing agreements with Antioqueños, and their success also usually translated into educating their children – generally lighter-skinned than themselves due to their 'marrying up' – in the white–mestizo cities of the interior. Success for black people thus seemed to lead to increasing connections with a non-black world. Overall, success and progress seemed to be structurally linked to whiteness or whitening, with that link being reinforced by Antioqueños' relative, but not absolute, exclusion of Chocoanos; and by the choice of some successful Chocoanos to 'marry up'.

Big-city Colombia

A parallel scenario occurred when migrants from the Chocó went to work and live in Medellín, capital city of Antioquia (Wade 1993). Here they formed a very small black minority in an overwhelmingly non-black, and indeed relatively white, city. Not surprisingly, given the regional background they came from, three-quarters of Chocoanos occupied the lower echelons of the job market: construction for men, domestic service for women and informal income-earning opportunities centred around selling food on the streets. A few worked as teachers and policemen, and there was a small Chocoano middle class of professionals (lawyers, doctors, etc.).

The fact that Chocoanos occupied a low position in the city economy is evidence of racial inequality, but not necessarily of racial discrimination in the job market. Their position could be due to their young average age, lack of qualifications and the simple fact of being migrants. Indeed, the Chocoanos were in many respects in a similar economic position to other, non-black migrants to the city of a similar age and educational status. Even when compared with these immediate non-black peers, however, there were some telling differences. Black women were concentrated more heavily in domestic service and black men in construction, both unstable and low-paying occupations. Was this a result of racial discrimination in the job market?

It is not easy to say. These concentrations could be due to the way personal networks operate to connect people with job opportunities: once a small number of Chocoano women found work as domestic servants, they found similar work in the city for their relatives and friends from back home. But the heavy overrepresentation of Chocoano women in domestic service, even compared to other young, poorly educated, non-black migrants, strongly suggests that they found trouble getting work in other occupations. Anecdotal evidence from young black women in the city confirmed that Antioqueños readily assumed they were domestics, for example asking Chocoano university students in the street if they were looking for work as a maid. Also, while Chocoano men and women left the Chocó for Antioquia in equal numbers, the women outnumbered the men in the city, which suggests that the men found it hard to get work there and had to find work in rural areas as unskilled labourers.

I found some evidence of clear discrimination. One woman recounted that an employment agency had simply told her, 'We don't place blacks here.' She said she felt as if a bucket of cold water had been thrown over her, but she did not link it to a more generalised pattern of racism in the city; it was a one-off incident, in her view. Others had similar tales of particular incidents, again usually seen as the bad behaviour of individual racists.

More systematic evidence came from the housing market. As expected, given their economic position, most Chocoanos lived in relatively poor accommodation, alongside many other non-black migrants and city-dwellers, in the peripheral *barrios* (neighbourhoods) and including in some precarious illegal or semi-legal invasion settlements. There was little evidence of overall residential segregation by race. However, two factors indicated the presence of racism in the housing market. First, I conducted an experiment in the room-rental market – which is a particularly sensitive area, as people are mostly renting out rooms in their private houses. This involved sending black and non-black people to rent a random sample of rooms to let, collected from the classified advertisements. It was clear that black applicants suffered a significantly higher rate of rejection than the non-black ones.

Second, for reasons associated with the way migrants find housing in rapidly urbanising cities, small nuclei of Chocoano settlement formed in specific *barrios*, usually quite precarious ones. Information would spread from initial settlers to friends and relatives, who would seek housing in the same area. Such nuclei often housed makeshift weekend dance halls, usually based in someone's house, where Chocoano migrants would come to party and dance to their preferred diet of salsa and *vallenato* accordion music from Colombia's Caribbean coastal region, both played at high volume. This fostered the consolidation of the nucleus of settlement. However, such nuclei – and especially the partying and loud music – attracted the ire of local Antioqueño residents, who sometimes harassed the Chocoanos and occasionally picked fights. Violence aside, the Antioqueños disapproved of such behaviour, which they saw as disreputable, sexually licentious, disorganised, unneighbourly and, in their view, typical of *los negros* and their supposed obsession with partying and sex, rather than the business of work and getting ahead in life.

Such views could also be found in relation to the Sunday city-centre gatherings of young Chocoano men and women (many of whom were domestic servants on their day off): one local newspaper reported that the Chocó had 'inundated our main recreational centres in Medellín' (see Fig. 6.1), and referred to the 'bad habits, excesses and lack of culture and the vulgarity of some Chocoanos … with no curb on their instincts'. As in Unguía, the Antioqueños deployed deeply rooted stereotypes about black people as lacking in moral fibre, preoccupied with sex, disorganised and not interested in progress. These views came out in specific contexts of neighbourhood confrontations and frictions, but what is striking is the consistent use of these stereotypes in situations as varied as Unguía and Medellín.

Figure 6.1. Afro-Colombian migrants in a Medellín city-centre bar.
(Photo by Peter Wade)

that low social class to begin with. Thus in Cali black people (migrant or not) had poorer educational status. They tended to find their first job more often in unskilled manual occupations than did white and mestizo people (72 per cent compared to 52 per cent) (Viáfara López 2008).

This is largely a result of historical patterns that, from colonial times, created a society in which black, indigenous and darker-skinned people were located in the lower echelons of the social hierarchy. Such hierarchical systems may permit some upward mobility, but they also have mechanisms that tend to reproduce the status quo, as valued resources (capital, education, skills, networks) pass preferentially from one generation to another within the classes that control those resources. People at the bottom of the system will, by definition, encounter barriers to their upward movement, because they have limited access to those resources. In the United States, for example, nearly 50 per cent of the advantages of having a high income are passed on to children; over 40 per cent of those whose parents were in the bottom income-earning quintile remained in that quintile. The more unequal the society, the more difficult the upward mobility.

If black and indigenous people were highly concentrated in the lower echelons of the hierarchy early on, then, even if there were no direct racial discrimination at work, it would take a long time for these racial categories to become evenly distributed across the hierarchy, especially in a highly unequal society such as Colombia or Brazil. (Just *how long* it should take is a vital but unanswered question.) If there is also racial discrimination at work – as we know there is – then full racial equality will take even longer or be permanently blocked. If there is some tendency (even a minority one) for upwardly mobile black and indigenous people to 'marry up' and have lighter-skinned children – as we know they sometimes do – then this will also impede black and indigenous categories from spreading equally across the social hierarchy.

In the meantime, in Colombia in particular there are processes affecting whole regions, rather than social classes, which exacerbate these effects for black and indigenous people. Such historical processes have resulted in the underdevelopment of the whole Pacific coastal region, the 'black region' of Colombia (but which also has important indigenous groups), and the less marked but still significant underdevelopment of large areas of the Caribbean coastal region, where over 30% of Afro-Colombians live, according to the 2005 census. The same census shows that people with 'unsatisfied basic needs' – a government measure of poverty – comprise 28% of the national population, but 80% of the population of the Chocó, about 45–50% of the provinces of the Caribbean coastal region and about 43% for all municipalities with an Afro-Colombian population of 30% or more. The data show that the infant mortality rate for Afro-Colombian males is over 48 compared to 27 for the Colombian population as a whole (Rodríguez Garavito, Alfonso Sierra and Cavelier Adarve 2009).

All this means that blackness (and indigeneity) is *structurally associated* with poverty and low social status in Colombia in ways that are deeply rooted and hard to shift. As we have seen, even apparently progressive reforms resulting in greater recognition of black and indigenous minorities and some material benefits for them have not been able to shift that association, which remains deeply entrenched or even exacerbated by recent processes of displacement. The structural association means that black people tend to enter the markets in a disadvantageous position to begin with. It then feeds direct processes of discrimination in the marketplace, as people who control resources limit the access of black people to housing and jobs – albeit not in strongly segregationist ways – because of the stereotyped link between blackness and poverty. This reinforces the association and a vicious circle is created.

The link between poverty and non-whiteness is only part of the story, however, as we have seen. Stereotypes of Chocoanos and Antioqueños are much more elaborate than this: in the dominant ideology whiteness is associated with moral qualities of hard work, sexual propriety, progressiveness, modernity and beauty, while blackness is associated with disreputability, sexual excess, vulgarity, laziness and ugliness. These stereotypes are not uniform: in Unguía and Medellín the Antioqueños were seen by the Chocoanos as rapacious, exploitative, potentially violent and possibly racist. Blackness (and brownness) can also be associated in the dominant ideology with positive values, such as fun, authenticity, sexual liberation and peace. There is a whole moral discourse around race, which goes beyond just poverty and wealth and encompasses people as cultural beings located in a hierarchy not just of wealth but of cultural value. Economic value and cultural worth are tightly interwoven, even if in varied ways (such that 'wealth' can mean 'arrogance' as well as 'modernity').

This discourse encompasses the nation as a whole. The presence of Antioqueños in Unguía was at one level about individual people building successful businesses and lives in a frontier region. But at the level of the nation it was also about whiter people bringing economic and cultural progress and modernity to an underdeveloped region, traditionally inhabited by darker-skinned people. Likewise, when Chocoanos migrate to Medellín or Cali it is not just about people relocating to find work, but about moving up the national scale of progress, modernity and cultural value, symbolised by the big, non-black city.

Racism and *mestizaje*

These processes of racial discrimination and racial inequality, which differentiate between white and black (and indigenous) as social positions and identities, are always mediated by processes of *mestizaje*, which blur this differentiation. This coexistence of separation and mixture is vital to understanding racial dynamics in Colombia and in Latin America. These are interdependent processes, because mixture depends on the existence of difference. But mixture blurs difference by interposing intermediate terms, even if these gain their significance from the differences that originated them. This blurring has been seen by some in Latin America to actually displace difference, such that *mestizaje* has on occasion been held up as a Latin American antidote to racism: how, it is asked, can there be racism when 'everyone is mixed'? (Wade 2004). This is a misrecognition – one crafted in the interests of presenting an image of 'racial democracy' to the world – but it is based on a seed of truth, which makes it persuasive.

We saw that, in Unguía, Chocoanos and Costenōs and even some Antioqueños mixed together at many levels – in families, in school, in dance halls, in the town. Successful Chocoanos tended to marry light-skinned partners and send their children to university in the interior of the country, where some of them partnered with lighter-skinned partners.

In Medellín, Chocoanos were a very small minority. Some of them lived in black nuclei, at least initially, but most dispersed around the city and lived in non-black neighbour-hoods; their children acquired Antioqueño accents, went to school with Antioqueños, befriended them and married them. In Cali similar patterns occurred, and lighter skin was associated with higher economic position: statistical data that distinguished *negros* from *mulatos* in phenotypical terms showed clearly that *mulatos* were higher than *negros* in terms of educational and occupational status, suggesting real processes of assimilation (Viáfara López 2008).

Mulatos appear here as a category for statistical purposes, but their mention attests to the more general fact that *mestizaje* inevitably blurs categorical distinctions of race. In Unguía the presence of the Costeños in particular made such distinctions rather inde-terminate. In Cali, where there has been a resident black population since colonial times as well as extensive black immigration, the categories of *negro, moreno, mulato*, mestizo and *blanco* overlap a good deal and are quite subjective. Discriminating against 'black' people or feeling discriminated against by a 'mestizo' were not processes that defined clear categories of people. The point is that all of this was going on *simultaneously* with stereotyping, discrimination, turning away black renters, assuming young black women were maids, violent harassment of black nuclei in Medellín, the desire of some successful black people to 'whiten', and so on.

This simultaneity of racism and mixture means the former is unsystematic, hard to pin down, always subject to counter-examples and easy to deny. The very things that are taken as evidence of racism can also be taken as evidence of its absence, depending on one's point of view. The fact that within a single family there may be people who are called by different colour terms (*moreno, güero*) suggests the absence of racism, but is also a location for its operation, as the darker-skinned family member may feel 'slighted' (see Chapter 5). The fact that darker and lighter people marry is evidence of racial democ-racy, but also a site of racism, because people's choices of partner may be judged, if not by themselves then by others, in racial terms (e.g. as being likely to produce 'prettier' or 'uglier' children, or as betraying an assumed desire among some black people to whiten themselves and their children). The fact that racial identities are ambiguous and subjec-tive is evidence of the irrelevance of race, yet also of its relevance, as people still make judgements based on their perceptions of race, however subjective these may be.

Just as important is that *mestizaje* operates partly in the domains of the family and sexual relations. This gives it both very public and very personal dimensions. The idea of the nation as having been constituted by the sexual encounter between Europeans, Africans and native Americans is a public narrative; the image of the mestizo nation, with its often minority black and indigenous populations, is also a public one. The partner choices people make and how light or dark their children turn out (see Chapter 5) are public matters too. Yet they are all grounded in personal relationships of sex, gender and kinship, and this gives *mestizaje*, in all its ambivalent connotations of racism and racial democracy, a powerful traction in people's lives: these ideas and discourses operate in the fabric of everyday life at a fundamental level. Ideas about mixture are not just a national-ist ideology of assimilation, but an experience lived through intimate aspects of social life.

DOMESTIC SERVICE

Domestic service is a useful example of how these lessons from Colombia work together. In much of Latin America female domestics are often black, indigenous or dark-skinned mestizo women serving middle-class families who are generally lighter skinned than they are. Black and brown women are overrepresented in domestic service in countries such as Colombia and Brazil: 2009 data for Brazil showed that 22 per cent of black women worked as domestics compared to 13 per cent of white women. In Andean countries such as Peru, domestic service is a very frequent occupation for young indigenous and dark-skinned females who migrate from the highlands to work in middle-class households in the lowland cities. This concentration in domestic service responds to both the structural position of young, non-white women – their low economic status and poor education – and processes of discrimination in the urban marketplace, where people assume that such women are domestics and other opportunities are less open to them.

But the role of domestic service is also linked strongly to processes of mixture and modernisation. *Mestizaje* as an ideology always involves the foundational story of white men having sex with indigenous and black women. Such women and their mestizo offspring were always the service class from early colonial times. At least some of the sexual encounters involved in *mestizaje* would have taken place precisely in these intimate semi-familial contexts of domestic service, where young women were highly vulnerable. The same continues to be true even today, as young domestics often report sexual harassment in the households where they work and the idea that young middle-class males might have their first sexual experience with the domestic servant is quite widespread. At the same time, the young migrant woman who works in an urban middle-class household is also seen – and often sees herself – as on the road leading away from rurality and indigeneity or blackness, towards urbanity, mixedness and whiteness. Thus in the Andes the term *chola*, generally used to mean an urbanised or acculturated indigenous woman, is often used in middle-class households as a synonym for 'servant'.

The typical servant is a young dark-skinned woman who 'offers herself' for service and thus for cultural 'improvement', implying distance from blackness and indigeneity, and suggesting (in the minds of some household men) availability to participate in sexual processes of *mestizaje*. These meanings around domestic service point to the cultural values attached to race, which are not just about low economic status, but about moral worth too. They point to the simultaneity of racism and *mestizaje*, as domestic service is a site for both processes, locating young, dark-skinned women as socially subservient, as sexual objects and as candidates for cultural whitening. The gendered and sexualised character of *mestizaje* is laid bare (Radcliffe 1990; Ribeiro Corossacz 2015; Saldaña-Tejeda 2012; Wade 2013a).

*

In Colombia a controversy, which illustrated some of these meanings, emerged when a Spanish celebrity magazine, ¡Hola!, published an article in December 2011 about the women of an elite family in Cali. A photograph showed the four women seated on a sofa, with two black maids posed behind them, in profile, dressed in white uniforms and holding trays.

- Some social scientists accused the magazine of racism: the photo glorified the structural subordination of black women and revelled in the privilege of whiteness.
- The photographer defended the photo as a simple depiction of reality: rich people in Cali hired maids; they happened to be black; so what? Indeed, the fact that they were in the picture at all was because they were loved and included in the family.
- A men's magazine countered by publishing a cover photo of four Afro-Colombian models in a luxurious setting, with two white maids posed behind them, holding trays – except that the four black women were completely naked.

This illustrates how the same item could figure as evidence both of racism and of an inclusive normality. Meanwhile, the satirical critique of the photo reversed its class connotations, but reinforced the sexist meanings of domestic service by blatantly sexualising the black models (233grados.com 2012).

6.5 Brazil: variations on a theme

Brazil has been a focus of social science studies on race for decades, and has a very rich literature on the topic (Hanchard 1999; Reichmann 1999; Reiter and Mitchell 2010; Telles 2004), including a good deal of ethnographic work. Many of the conclusions from the Colombian material work broadly for Brazil as well – the awkwardness of race as a topic, the unsystematic character of discrimination, the association between blackness and low status, the value often attached to whiteness, the tendency for some successful black people to 'marry up' racially, the important but ambivalent results of black social mobilisation and state recognition of it, the coexistence of racism and mixture, of racism and its denial, and the gendered workings of *mestizaje* (see, for example, Burdick 1998; Caldwell 2007; Fry 2000; Goldstein 2003; Sansone 2003; Sheriff 2001; Twine 1998). But there are certain key variations in the way racial differences and inequalities are lived and reproduced (Wade 2013c).

Demographics and statistics

Brazil is one of the few Latin American countries to have included a census colour question over a long period of time in the twentieth century. This has generated data that can be used to address in detail questions of racial demography and inequality. Compared to Colombia, Brazil has a much larger white population (48 per cent of people self-identified as such in 2010), due mainly to large-scale European immigration. Whiteness dominates regions of the south, where whites are 70–85 per cent, but no region is predominantly black in the way blackness dominates the Pacific coastal region of Colombia – although parts of the north and north-east are predominantly *pardo* (brown, a very ambiguous and inclusive category). This white presence has contributed to the possibility, nowadays routinely used in state statistics and black movement discourse, of dividing the Brazilian population into 'whites' and 'blacks' (*negros* = all *pretos* and *pardos* in census terms). Statistically this is seen as justified because although *pretos* are slightly below *pardos* in terms of average socio-economic profile, they are both distinguishable from whites. In Colombia when a binary division is made for statistical and political purposes it is between 'non-black' (all whites and mestizos) and 'black' (probably close to the Brazilian *preto*).

Official statistics have been used by social scientists in Brazil to demonstrate more comprehensively than for Colombia some key points:

* Racial inequality is marked in education, the job market and health outcomes: non-whites' illiteracy rate is double that of whites, while they earn half as much. Black women die in childbirth seven times more often than white women (Bailey 2009; Burdick 2013: 6).
* Racial discrimination does operate, and racial inequality cannot be explained by initial black disadvantage simply being reproduced by mechanisms of class inequality. One study concluded that 24 per cent of the wage-earning difference between white men

and black men was likely due to racial discrimination, independently of class differences (Lovell 1994). (As a comparison, the same study concluded that sex discrimination accounted for 85 per cent of the income differences between white men and white women.)

- Residential racial segregation is more marked in Brazil than in Cali (the only Colombian city where it has been measured). The black/white 'dissimilarity index' – a measure of residential segregation in which 100 indicates complete segregation of two populations and 0 represents complete integration – was 48 for the city of Salvador, in the north-east, in 1980, compared to 29 for Cali in 1998. This is only about half that of Chicago in 1980. The data show that this segregation is a little stronger for the middle classes (Barbary 2004: 187; Telles 2004: 203, 209).

Difference and *mestiçagem*: the question of affirmative action

In Brazil this black–white opposition, used in particular by the state, statisticians and the black movement, sits alongside a more flexible set of everyday racial classifications which, although they can involve dozens of descriptive colour terms, in practice tend to centre on *branco* and *moreno*, with *pardo, preto, negro* and *moreno claro* (light brown) being used with less frequency (see Chapter 5). Everyday classifications also vary somewhat depending on who is making them: dwellers in Rio's *favela*s (low-income settlements or 'shanty towns'), for example, alongside the use of several colour terms, do recognise a basic division between white and black, while middle- and upper-class people tend to emphasise fluidity and mixture (Sheriff 2003).

These different classificatory tendencies speak to the simultaneous coexistence of racial difference and racial mixture, and to the possibility of emphasising one or the other. This is a common feature in Latin America, but it takes on a specific shape in Brazil, because (a) the ideology of mixedness as the basis of national identity was developed with particular force from the 1930s and celebrated through cultural icons such as samba, seen as a highly syncretic music with strong black roots; while at the same time, (b) the socio-economic and demographic realities of the country laid the basis for perceiving society as divided into white and non-white categories.

Nowhere are these tensions more clearly visible than in the heated debates over race-based affirmative actions (see Chapter 4). In Brazil, more than in any other Latin American country, the existence of hard data on racial inequality, plus the possibility of thinking in terms of apparently well-defined categories of people, used to self-identifying by colour or race, led to the implementation of affirmative action as part of multiculturalist reforms. In Colombia the 1993 law targeted affirmative actions at rural black communities in one region; only later did the state begin to address blackness as a racial category cross-cutting rural and urban contexts. In contrast, in Brazil blackness has long had this broader racial and urban dimension, and so affirmative actions targeted black people in general. A key plank in this programme was the introduction of admissions quotas for 'black' (*negro*) students in some public universities. Quotas were also made available for students from public schools, but these did not cause much controversy: it was the racial quotas that generated debate. (In Brazil public secondary schools are considered poor, but public universities are valued.) The discussions centred around whether it was appropriate – in general and specifically in Brazil – to attempt to correct racial inequality by introducing race-based measures (Fry 2000; Htun 2004; Skidmore 2003).

Those against the idea argued that, ideally, the aim was to erase racial distinctions, whereas these policies reinforced such distinctions by enshrining them in legal rights. This camp did not deny that racial inequality and racism existed, but contested these policies as the best way to manage them. It was better to address all social inequalities together and introduce reforms into the education system that would benefit all poor people, including blacks. Also, these race-based policies made even less sense in Brazil, because racial identities and categories were anyway diffuse and uncertain, so it was not practical to define a racial category of beneficiaries; such a definition, if imposed, would simply harden racial divisions, rather than erode them. Many were shocked at the use in the University of Brasilia of a committee to establish who was 'black' (most universities sensibly opted for self-identification). Some saw the policies as reflecting a US-style approach to race that did not suit Brazil.

Those in favour of the policies said that racism and racial inequality needed targeted measures to address them. Implementing social reform benefiting all the poor was a good idea, but could never address the persistence of racism, which operated within the class system to disadvantage blacks in particular ways. Race-based actions had the potential to rapidly accelerate the upward mobility of some black people and seriously alter the racial profile of the middle classes, disturbing the powerful and debilitating structural association between blackness and low status. Affirmative action policies still exist, despite legal challenges, but the debates about them clearly show the tensions between racial difference (blacks versus whites) and racial mixture ('we are all mixed') in ideas about the character of the Brazilian nation (see Fig. 6.2).

Racial quotas in practice

Such debates took place at the level of theory and politics, but some anthropologists also did university ethnographies to see how racial quotas worked in practice. André Cicalo (2012a, 2012b) focused on law students in UERJ (State University of Rio de Janeiro). The racial quotas had changed the face of the law programme, which previously had been predominantly white and was now more mixed. Cicalo worked mainly with students who, although they were aware of the Brazilian black social movement, which had been active in lobbying for quotas and the use of the *negro* category to include *pretos* and *pardos*, were not themselves militants. Reflecting practices in the wider society, he found that race and racism were not subjects that people immediately referred to in everyday interaction: there was a certain silence about race, despite the very public status of the debate about racial quotas. But race, always intertwined with class and often gender too, surfaced in more subtle ways. White and black female students admitted feeling nervous about street encounters with young black men: one visitor to Cicalo's apartment got a shock when she met with *um negão* (a big black man) in the hallway of the middle-class building. He was a

Figure 6.2. Students at São Paulo University supporting racial quotas. The sign reads, 'Those against quotas are racists.'
(© Marcelo Camargo/ABr)

resident, but was, for her, an unexpected person in that context – in contrast to the many black females, who were maids in the apartments.

The connection of race and class, especially in a mainly elite law programme, was a constant theme: quota students were generally assumed to be dark skinned and vice versa, despite the fact that some students admitted through public-school quotas were not dark skinned and that some black students were not quota students. Quota students tended to congregate in the morning schedules, as they often worked in the evenings. Within the classroom there was a tendency, not sharp but noticeable, for the 'swots' to sit in the front, while more relaxed students sat at the back. Quota students, less confident about their abilities and eager to learn, tended to be labelled 'swots' or 'nerds', while confident white students from good schools sat at the back, some of them forming a social clique called the Barbarians. This was not clear-cut: there were some poor white students and a few middle-class blacks too. But a couple of white quota students were able to socialise with the Barbarians with little comment, while one better-off black man, who was seen by the nerds as the 'Barbarians' black', in the end couldn't keep up with their expensive social life.

The consumption habits of the richer, whiter students constantly separated them from the poorer, darker ones: the former wore fashionable clothing that looked informal but cost a lot; the latter wore clothes that looked cheaper. When the class went on a trip to a courtroom situated in a poor part of town, it was clear that many white students had never been in such a neighbourhood, nor even on a suburban train; on the way back the darker-skinned students got off the train at nearby stations, while the white students continued towards the richer, whiter neighbourhoods. On the other hand, some middle-class white students said they enjoyed the fact that the university was now more mixed, a bit 'rougher' and less elitist than in the past; they felt this made it more authentic in Brazilian terms.

Still, the wealthier students tended to socialise in the law faculty's Athletic Society, traditionally a white and elite space, while the poorer students hung out in the faculty students' union. Interestingly, one of the few very overt references to race came in the context of the Athletic Society's annual games, a quite costly and hence predominantly white event at which students from a rival elite private Catholic university began jokingly to refer to the UERJ people as 'Congo', alluding to the darker profile of the university post-quotas. Soon, the Barbarians appropriated the epithet, adopting a Congolese national flag and shouting 'Yes, we're Congo ... [and] we have a [big] black dick!' Blackness could be appropriated by whites as a kind of sexual authenticity (Cicalo 2012b: 88).

Cicalo found that students' identities were not transformed in a straightforward way into simple, well-defined categories of black and white, despite an admissions process that demanded such clarity. While militant black students often operated with a clear black–white binary, at least in some contexts, others used more complex modes of reckoning. Schwartzman (2009) found that, in deciding who they were in racial terms, students juggled with various modes of categorisation and identification, none of which formed fully coherent systems: they were familiar with census categories (including the term *pardo*, which they rarely used outside this kind of bureaucratic context), with the state and black movement black–white binary, and with the more flexible categorisations of everyday life. Individual students would juggle with elements of all these, also working out whether they should, in ethical terms, identify as *negro* for quota purposes. Were quotas designed for all *negros* or really for people who suffered racial discrimination (which could be a subset of that broad category)? Had they ever suffered racial discrimination themselves? The question required some soul-searching. Or could racism be deduced from observing racial stratification in Brazil, independent of their own experiences?

Overall, then, racial quotas did not have the effect of simply transforming students' racial identifications. Rather, binary racial difference continued to coexist with ideas about racial mixture. Still, Cicalo observed that, for black students from a poor background, the experience of going to UERJ was a transformative one. UERJ was a material and symbolic space of dominant white society, set up to train people to lead and govern the country. Admission to this space gave these students a different perspective on their place in society. At university they also encountered more white middle-class people, at least in theory on equal terms. They also, on average and after a settling-in period, competed well with others in terms of grades (thus confounding one criticism of quotas, which held that they depressed academic standards). And they were more exposed to the ideas of the black militants. All this led to a more critical attitude and more self-confidence, and this might be expressed, for some people, in a more open expression of black identity – for example, one woman began to wear braided hair and frequent clubs playing 'black' music that she had previously avoided.

Lessons from Brazil

In the context of Latin America the presence of race in Brazil is rather explicit, in the sense that the census enumerates by race/colour, the statistics suggest a white–black divide, and iconic cultural forms, such as samba and carnival, are associated with black origins. Recent reforms have made this presence more palpable and consolidated the idea of a white–black racial binary. Yet the presence of mixture is equally powerful in the country, in the sense that the idea of racial democracy has had its strongest expression there and the image of the nation as a fundamentally mixed one, racially and culturally, has been elaborated powerfully and in great detail. Recent multiculturalist policies have not undermined this sense of mixture, indicating that the tension between racial difference and mixture still lies at the heart of Brazilian and, more widely, Latin American racial orders and that ideas and practices of mixture can still create a context in which race and racism are still awkward topics. Meanwhile, affirmative action is unquestionably helping to change, albeit slightly, the racial profile of the middle class. For the moment, however, the structural association between blackness and low status remains, and penetrates even into the arenas where the black middle class is being created.

6.6 Guatemala: racial ambivalence

Although Colombia and Brazil have indigenous populations, and a full understanding of race in each country would require considering indigeneity alongside blackness, it makes sense to explore racial difference and inequality with respect to indigenous populations in contexts where indigenous peoples form the main counterpoint to white and mestizo categories. Guatemala is an interesting case, because the boundary between *ladinos* (the local term for a non-indigenous person) and mainly Maya indigenous people is quite strongly marked, compared to the country's neighbour Mexico (see Chapter 5). In the 1950s some US anthropologists talked about the boundary in terms of 'caste', a term derived from India, which was also being used in the United States to describe black–white relations under the highly segregated 'Jim Crow' conditions of the US South. (Even then, however, the boundary was described in terms of culture not race.) The boundary was reinforced by the military repression of the 1960s through the 1990s, when indigenous Maya communities were targeted as subversive collaborators of left-wing guerrilla movements.

Charles Hale's ethnography of the town of Chimaltenango gives great insight into the changing ways racial difference and racial inequality are generated in Guatemala. Hale notes that in the 1950s and 1960s, 'if an Indian presumed to share the sidewalk with *ladinos*, he was fair game for verbal and physical abuse' (Hale 2006: 54). *Ladinos* dominated local politics and landholding (they were 23 per cent of landowners, but owned over 60 per cent of the land). *Ladinos* and Maya people did not interact on equal terms in any social domain, and the two groups were mostly set apart. Indigenous people did form cooperatives, try to access international development funding, and organise to challenge inequality, but the military state harshly repressed indigenous mobilisations. An indigenous man, who in 1959 was elected to mayor in a small locality where 89 per cent of landowners were indigenous, was forced to resign by threats that made him fear 'the murder of his race'. Although some *ladinos* were left wing and anti-racist, many espoused frankly racist views, sometimes with a marked discourse of biological difference.

Already before the 1990s structural patterns were beginning to change. Indigenous people had begun to organise more widely, despite a combination of genocidal repression and 'disciplinary assimilation', and had formed many Maya NGOs; some of these had access to international funding, which favoured indigenous people. Maya voters had also achieved local-level electoral victories in some areas. A small educated indigenous middle class had begun to form, taking some positions in the public and private sectors. Indigenous people were no longer so deferential, and some of them were interacting with *ladinos* on more equal terms. A key figure in the Maya movement, Rigoberta Menchú, achieved international fame in the 1980s when her biography describing army atrocities was published in several languages; she was awarded the Nobel Peace Prize in 1992.

After the end of the civil war in 1996, and in line with the multicultural reforms sweeping Latin America in the 1990s, *ladinos* began to adopt a posture of 'racial ambivalence'. On the one hand, they mostly decried the racism of the past, declared their support for multiculturalism and equality, and spoke in terms of cultural differences, rather than anything remotely biological; on the other hand, many of them were unwilling to relinquish the material and symbolic privileges of the *ladino* position.

Ladino dominance remained in many respects: they still controlled the lion's share of the economy and the political domain, despite indigenous challenges. But the means of reproducing racial privilege changed somewhat. Hale argues that the state adopted a more progressive stance on Maya cultural rights, but in a limited way, by incorporating them into the state apparatus – often in areas where Maya people had anyway begun to make advances (e.g. in local politics and in the public administration) – as long as they proved cooperative and not overly radical and demanding. Hale labels this 'the *indio permitido* (the authorised Indian)', who is allowed in, providing he or she behaves in a way deemed acceptable by those in power.

Hale links this conditional recognition of indigenous rights to neoliberal reforms. These are in part economic measures designed to liberalise markets for the free movement of money and labour, and scale back state commitments to social welfare, but they also include measures aimed at reforming political governance, such that not just individuals but also groups become harnessed to the state and the market as self-regulating and ideally entrepreneurial entities, who will support the development of capitalist enterprise. 'Neoliberal multiculturalism' means that ethnic minorities are cast as supporting players in a large-scale project of economic and political reform designed to further capital accumulation.

Ladinos in Chimaltenango felt uncomfortable with this new state of affairs, which threatened their sense of social position. They tended to react by accusing Maya people of 'reverse racism', an accusation one can hear occasionally in Colombia and Brazil, made by whites and mestizos against black and indigenous activists. The idea is that the *ladinos* have given up their racism, accepted the error of their ways and are willing to deal with

Maya on equal terms – and Hale makes clear that, for most *ladinos*, this is not just a cynical pose, but a sincere commitment. But, they say, the Mayas insist on maintaining a sense of separation, seeing *ladinos* as different (even inferior), and using that difference to their advantage (e.g. to claim opportunities linked to international funding or the recognition of indigenous groups). The *ladinos* hark back insistently to an indigenous uprising of the 1940s, which they recount as an attempt by the Mayas to massacre *ladinos*, and they recount incidents from the more recent history of Maya electoral success as evidence of an indigenous desire for revenge.

The 'insurrectionary Indian' is a powerful figure in the local *ladino* imaginary, and is not based on actual indigenous violence in recent times, but on these older histories, which have assumed mythical proportions. The image is fuelled by sexual relations between some middle-class indigenous men and *ladino* women, which are often assumed to be driven by the men's desire to avenge themselves for *ladino* sexual predation on indigenous women in previous times (see also Nelson 1999: ch. 6). *Ladinos* see themselves as the nation's natural guardians and leaders, and see assertive Mayas as usurping this role by making a claim to being the original and authentic population. Meanwhile, they argue, Maya activists are themselves not authentically indigenous, because they are educated and wear sports shoes and jeans, while the average rural Maya actually wants their children to learn Spanish in school and get ahead in the modern world. In short, *ladino* discourse is shot through with an image of themselves as the victims of a racism enacted by indigenous activists falsely claiming an authentic identity and pretending to represent grass-roots people, who themselves have no interest in racial authenticity.

JOKING ASIDE

Among Guatemalan *ladinos*, Rigoberta Menchú was the butt of many jokes that reflected the darker side of racial ambivalence (Nelson 1999: 376):

- Mattel is now making a doll of Rigoberta, which makes Barbie and Ken happy because now they'll have a maid.
- Rigoberta will not invest the Nobel Prize money in buying a plantation because she's afraid the Indians will invade the land.
- I heard that Rigo is buying a house in Zone Fourteen [an elite neighbourhood in the capital city] and she's put out word that she's looking for a *muchacha* [a girl, a maid] among the *ladino* girls who live there.

These jokes cast Menchú in contradictory ways as (a) still really a maid; (b) an inauthentic Maya, who is, like *ladinos*, afraid of indigenous invaders; and (c) a reverse racist who wants to dominate *ladinos*.

Guatemala, like many other regions of Latin America and the world, has undergone a shift – albeit belatedly – towards a discourse of anti-racism and cultural, not racial, difference. As in many other regions, ideas about race persist, not always under that exact name, but certainly recognisable as such: in Guatemala race remains very visible. Hale shows how even progressive anti-racist *ladinos* will sometimes indulge in kitchen-table conversations in which they discuss the supposed physical, bodily differences between *ladinos* and indigenous people – noses, hips, thighs, skin colour (2006: 25). What stands out from the Guatemalan material – although not unique to Guatemala – is the way racial difference is expressed by *ladinos* through pervasive and anxious discourses about reverse racism, racial authenticity and the threat of the racial other, discourses mediated by Guatemala's particular history of racialised violence. The way such discourses of racial difference feed into practices that reproduce racial inequality is not explored in detail in Hale's ethnography, but clearly the anxieties, suspicions and fears that characterise *ladino* ideas about Maya are likely to influence the way they allow Maya people access to valued resources they control.

6.7 Performing and embodying race in the Andes

As we have already seen (Chapter 5), it has often been claimed that race in Latin America is a rather cultural affair, malleable and contextual. The possibility that an *indio* can become a mestizo, by changing the way he talks, dresses, where he lives etc., is often adduced as evidence of this. Although such shifts may be harder for women, who are often perceived, in the Andes anyway, to be 'more Indian' (De la Cadena 1995), domestic service in the city can be enough to turn an *india* into a *chola*, if not a *mestiza*.

This malleability is certainly an important feature of race in Latin America, but it has to be seen in context. First, there are limits to it, especially for blackness and whiteness: a typically 'white' person cannot really become 'black' and vice versa, whereas the difference between a *negro* and a *moreno*, or between a *blanco* and a mestizo, is very much a matter of judgement. Second, the fact that racial identity may be malleable and contextual does not mean it is not significant socially: in the Andes, although there are not clear divides between white, mestizo and indigenous, the images of white and indigenous act as magnetic poles that structure the field of racialised identifications, within which 'one is more or less white; more or less indian' (Canessa 2012: 7). Third – which is the point I want to expand on here – the malleability of race does not detract from its *materiality*, especially in the body.

Mary Weismantel's work on race in the Andes – mainly Ecuador – is useful for grasping this point. She notes that many people in the Andes talked about race as a physical reality not reducible to class or culture: indigenous, white and mestizo people were physically different from each other. And yet people also 'spoke matter-of-factly about neighbours who had changed their race during their lifetime' (2001: 191). This apparent contradiction is not to be reconciled either by saying that race is not the issue here, because race by definition involves fixity, or by saying that race must in fact be just a matter of culture in this region. Instead, race has to be conceived as rooted partly in bodily processes of accumulation, transformation and change, processes that are performed by, and that also create bodies.

This is related to the environmentalist understandings of race we encountered in Chapters 2 and 3: eating different types of food, for example, is seen by Andeans as a vital component of what makes people indigenous or white, and can change them in one direction or another. Eating food is a gesture of solidarity, but it also forms the body in predictable directions, making them more or less white or indigenous. People's feet also bear the imprint of their environment and indicate their race: indigenous people who walk barefoot have splayed, calloused feet, completely different from those of people who have worn shoes all their life. Indigenous and white bodies are also assumed to give off different smells, as an indication of their distinctive constitutions. Racial indicators accumulate in the body.

Race is also performed in less bodily engrained ways, related to objects. Indigenous feet can be covered with shoes, and this can be interpreted in different ways. Whites and mestizos may assume that the underlying indigenousness remains unchanged, merely covered with a thin veneer, seen as not really fitting. Local Andeans react by saying that 'the body no longer belongs to the race': the shoe-wearers have become white and will also have to accustom their mouths to a new language, and their bodies to a new diet (Weismantel 2001: 190).

Behaviour is also important. For example, one figure that Weismantel explores is the mythical *pishtaco*, a kind of bogey-man, usually perceived as a white or mestizo, who steals indigenous people's body fat and sells it; he may also be a rapist. The *pishtaco* symbolises whiteness, dominance and rapaciousness: if an indigenous person starts acting in a fashion perceived as exploitative or domineering, s/he can be accused of being like a *pishtaco*, thus implicitly whitening – and indeed masculinising – him or her. The other main figure Weismantel presents is the Andean market woman, often labelled *chola* or *mestiza*: although they might be considered indigenous in many respects – and indeed do not deny or reject their indigenous origins – the fact that they work in urban environments, as traders, not farmers, and manage quite large amounts of money, which they carry round in fat rolls, concealed in their skirts, means they are not *indias*. They are racially ambiguous, and also sexually so: because of their money and assertiveness they are slightly masculinised, while also seen as sexually provocative. In De la Cadena's words (2000) they are 'indigenous mestizos', people who are indigenous, whose success and hard-working lifestyle makes them into mestizos – and who have few qualms about insulting those beneath them as *indios sucios* (dirty indians).

This Andean material teaches us that racial difference is performed or enacted – rather than just given permanently by fixed bodily features – but that such performances, while partly symbolic and related to objects (such as shoes), can also, through their constant reiteration, become materially engrained into the body in durable ways, without being immutable.

CONCLUSION

The main purpose of this chapter has been to explore how processes of racial differentiation and racial inequality operate in practice and in the detail of everyday behaviour. Latin America is a specific context in terms of the presence of *mestizaje* as an ideology and practice, the presence of mestizos as a category, the coexistence of elements of racism and 'racial democracy', the particularly malleable character of racial identities. But many of the processes involved are more widely relevant: the structural and historical roots of disadvantage, the self-perpetuating circles of inequality, the tactics of exclusion and inclusion, the denial of racism and its persistence as people follow their own interests, the embeddedness of race in other social processes (frontier expansion, city migration, finding somewhere to live, making a living, domestic service, national identity, left-wing politics) – these are all processes that find parallels in other regions where racial differentiation and racial inequality are reproduced. The useful thing about a focus on Latin America is that mixture and a cultural discourse of race are long-standing features that have become increasingly evident in other contexts – for example, the United States, Europe and South Africa. We can therefore see clearly how racial differentiation and racial inequality operate through and alongside these features.

I said at the beginning of the chapter that regions are interconnected and it is important to see that connections are always there. This is evident in particular ways – for example, how the black social movements in Colombia and Brazil drew inspiration from the US and South African black figureheads, or how the Guatemalan Mayan organisations could benefit from international funding from NGOs sympathetic to indigenous rights, or the way affirmative action in Brazil was seen by critics as being an inappropriate imposition onto Brazilian realities of a US model of race. And it is evident in more general ways, such as the way stereotypes about Chocoanos and Antioqueños in Unguía resonate with familiar images of blackness and whiteness that can be found throughout the Americas and more widely still in European ex-colonies.

FURTHER RESEARCH

1. Explore anthropological and other ethnographic work on specific countries, starting with the material referenced in this chapter (and in the section on Latin America in Chapter 5) and/or venturing into other sources, such as:

    a. Colombia: Streicker (1995); Asher (2009); Arocha (1998); Ng'weno (2007).

    b. Brazil: French (2009); Selka (2007); Pinho (2010); Warren (2001).

    c. Guatemala: Wilson (1995); Warren (1998); Pitarch, Speed and Leyva-Solano (2008); Sanford (2003).

    d. Andes: Larson, Harris and Tandeter (1995); Poole (1997); Roitman (2009); Whitten (1981); the journal special issue edited by Weismantel and Eisenman (1998), including articles by Orlove and Colloredo-Mansfeld.

    e. Mexico: Sue (2013); Lewis (2012).

2. If you read Spanish and/or Portuguese, use the sources already cited to lead you to the abundant material in these languages.

3. Use the text boxes on domestic service and on jokes about Rigoberta Menchú as starting points to research these particular topics.

4. Look at the representation of black women in the press coverage of the Cali incident reported in the text box on domestic service. Link this back to the Further Research item about images of Sara Baartman (Chapter 3).

7 The United States and South Africa: segregation and desegregation

Latin America and especially Brazil have often been seen as the polar opposite of the United States and South Africa in terms of racial formation. In the past Brazil touted itself as a racial democracy compared to the image of the United States as a racial hell; today, despite recognition that Brazil has its own problems of racism, some Brazilians still see race-based affirmative action as a US model of racial policy that is unsuited to their national reality, as we saw in the previous chapter. There are important differences between Latin America, the United States and South Africa, but these should not obscure the fact that they are all variants of a similar history of colonial rule, racial oppression and nation building with racial hierarchy at the core. The key differences lie in the way legal racial segregation and clearly defined racial identities came to characterise the United States and South Africa, and the legacies of these two features during the processes of formal desegregation that have occurred in the United States after WWII and in post-apartheid South Africa. In Latin America both features, while not absent, have been less marked.

In this chapter I explore why racial segregation emerged in the United States, how it operated and its continuing legacies today. As in the last chapter, I use ethnographies of particular places to illustrate the processes at work. At the end I deal briefly with South Africa, using it as a counterpoint to the United States, in order to examine segregation and desegregation in a context in which the subordinate black class is a native majority, rather than a deracinated minority, and has taken the reins of power.

7.1 Changing US demographics

In Chapter 2, in comparing patterns of race mixture and racial classification in the colonial Americas, we saw that the United States, especially in the South, developed along comparatively segregationist lines, drawing increasingly clear boundaries between white, black and native American from the late eighteenth century onwards, with black and slave being almost coterminous categories – although less so in the North – and an increasing emphasis on including mixed-race people in the subordinate categories. Table 7.1 shows how the country looked by 1860, just before the Civil War.

The table shows the huge concentration (92%) of black people in the South, where they were over a third of the total. Outside the South the country was overwhelmingly

TABLE 7.1 Racial demography of the USA, 1860 (percentages)

	White	Negro (including Free Coloured)	(Free Coloured)	Other Races	Total	n ('000s)
South	63	37	(2)	0.2	100%	11,133
	26	92	52	3		
Non-South	98	1.6	(1.1)	0.4	100%	20,310
	74	8	48	97		
USA total	86	14	(1.5)	0.2	100%	31,443
	100%	100%	100%	100%		
n ('000s)	26,922	4,441	488	78		

Source: US Bureau of the Census (1949: 25–7).

(98%) white. Free people were about 10% of the black population, but were overall a tiny minority (1.5%) everywhere. The percentage of black people in the total population has stayed fairly constant since then, but, especially after 1900, powerful processes of migration out of the South distributed black people all over the country; in some big Northern cities, such as Chicago and Philadelphia, they form 30–40% of the population, rising to over 80% in Detroit. However, in 2002, 55% of black people still lived in the South and in 2010 they formed 20% or more of the population only in the states of the South.

The black–white, North–South difference has been a crucial one in the history of the United States, but it has to be placed in relation to three other key axes: (a) white–Native American; (b) white–Mexican; and (c) white–Asian. In 1860 Native Americans, counted among 'other races', were a minute fraction of the total, but formed a larger proportion (10 per cent) in the Western states, at that time sparsely colonised by whites. Today American Indians are about 1 per cent of the US total population. Despite the small size of the population, they have long formed an important counterpoint to whiteness, being the native population displaced by white expansionism: the conquest of the western half of the country, mainly after the Civil War, involved a series of wars with Native Americans, during which their population was decimated. The figure of the 'Red Indian' was then, and remains today, a powerful symbol of racial difference.

The US–Mexico border was the site of an expansionist war in 1846–8, and has long been a crucial dimension of the US racial panorama, because it separated, albeit in an ambiguous way, the mainly white United States from Latin American neighbours to the south, perceived by white Americans as less white. The border was, and still is, a racialised as well as a territorial one. Long admitted as temporary workers, Mexicans and other Latin Americans have settled in greater numbers in recent decades, with Mexicans currently about half of all Latin American immigrants. Whether immigrant or US-born, people of 'Hispanic origin', to use the census term adopted in 1980, have risen from 2% of the total population in 1980 to some 15% in 2010: states where they form more than 20% of the population are in the south-west, plus Florida (US Bureau of the Census n.d.).

This situation is racially ambiguous because many Mexicans and Mexican Americans identify as white both in Mexico and in the United States. Mexicans figured as a specific 'race' only in the 1930 census and were thereafter counted as white, unless they were 'definitely Indian or of other Non-white races' (Rodríguez 2000: 212). Today about half of all Hispanics and Latino Americans identify as white, with over 40 per cent of them ticking the box for 'some other race' or 'two or more races' (Humes, Jones and Ramirez 2010). Despite this ambiguity, the US–Mexico border continues to mark a racialised difference between 'white' and 'brown'. Thus the increasing presence of Latinos in the United States has spurred a process of 'browning', which is blurring a simple black–white division.

Asians, who were mainly Chinese, were included in 'other races' in 1860. They were demographically a very small proportion overall, but formed about 10% of the Californian population in the late nineteenth century. Although a small minority, their presence spurred fears of the 'yellow peril' among whites, and the Chinese Exclusion Act was passed in 1882 to restrict immigration. By 1940 'other races' (now including various other Asian nationalities) were still less than 1% of the US total, rising through the 1970s and 1980s to form about 5–6% by 2010. Asians have accounted for much of this increase, forming about 30% of all immigrants by 2012, with East and South East Asians forming the majority (Migration Policy Institute 2014; US Bureau of the Census n.d.). Like Latinos, the Asian presence has contributed to a process of browning, which diffuses the black–white binary.

Building a segregated nation

The demographic sketch above is underlain by complex historical processes, in which race and nation figure large. One important element was the view that white Americans had of themselves and their nation in the nineteenth and early twentieth centuries, in relation to others, both within the nation and beyond. Horsman (1981) argues that many white Americans adhered to an ideology of 'Anglo-Saxonism', which held that Anglo-Saxons were a special breed of people, formed through a history of defending political and individual freedom in medieval England and who had successfully fought for freedom in the Americas against colonial rule. They were now allotted the historic role, by virtue of their racial superiority, of fulfilling their 'manifest destiny' and bringing these freedoms to their country as a whole and to other parts of the globe, albeit by the subjugation of peoples seen as inferior. Of course, many white Americans – even those of English descent – were not Anglo-Saxons, but whiteness served as an inclusive marker of Anglo-Saxonism and racial superiority. There were plenty of powerful economic reasons why white Americans wanted to conquer the western regions of the United States, take large swathes of territory from Mexico and colonise regions abroad, but the belief in this manifest destiny was a powerful justification for this mission and for the domination of peoples believed to be inferior and thus not as capable of sharing in the freedoms of political democracy.

A second element was the attempt to build a nation that had been terribly divided by a vicious civil war, which pitted North against South and slave states against non-slave states. Between 1861 and 1865 an estimated 750,000 men died in the conflict, more than in all the other major wars in which the United States has since been involved. Anthony

Marx (1998) highlights the role that nation building played in shaping racial formation. He argues that the need for post-war national unity, perceived as white unity, drove the move towards the intensive and strict racial segregation system, known as 'Jim Crow', which dominated in the South from the 1870s to the 1950s.

In the aftermath of the Civil War the North tried to impose 'Reconstruction' on the South, in the shape of extensive reforms obliging Southern states to fall into line with Northern policies, for example forcing Southern governments to accept black voting rights and a public school system open to blacks. This provoked Southern anger and opposition; the Ku Klux Klan started during this period, and political differences among Southern whites melted in the face of Northern reformism. Gradually, the North backed off, allowing Southern states greater autonomy and colluding in the disenfranchisement of blacks (by indirect means such as literacy tests, which also excluded poor white voters) and the imposition of a Jim Crow system in the South, in which laws forced the segregation of blacks and whites. (Jim Crow was a caricature black figure.) Such legal segregation was not imposed in the North, but de facto segregation occurred in housing and the job market.

Marx's contention is that the need to create some measure of national unity took precedence and that, in effect, whites of different regions and classes agreed to their common dominant position over blacks (and, one might add, other people considered non-white): the price of national (white) unity was black subordination. This led to a highly segregated system and consolidated an increasingly clear and encompassing definition of black, based on the 'one-drop rule', according to which anyone with 'one drop' of black blood was defined as black; in practice this meant anyone known to have black ancestors or whose appearance suggested black ancestry.

The role of class factors in all this is complex. Racial segregation did not straightforwardly serve the interests of white capitalists: they might benefit from a divided working class, but they also had to work with a labour force restricted by complex segregations. Nor did segregation simply serve the interests of the white working class, which could enjoy a feeling of superiority and even some material benefits vis-à-vis the black working class, but which might also suffer the effects of wages lowered by black–white competition. Instead, the racial formation of segregation reduced conflict in the nation – at the price of black oppression – and avoided major disruptions to the economic system as a whole. But segregation was driven as much by racial fears and ideas as by economic ones: the two aspects were intimately intertwined.

7.2 Caste and class in segregated Southern towns

A number of anthropological and sociological studies portrayed the character of segregation in the US South under Jim Crow. For example, John Dollard (1937) and Hortense Powdermaker (1939) both did ethnographic studies of the cotton town of Indianola, Mississippi, while Allison Davis, Burleigh Gardner and Mary Gardner (1941) carried out a study of Natchez, a cotton-belt town, also in Mississippi. They all described communities rigidly divided by race, into black and white 'castes' (Warner 1936) – with explicit, but vague, reference to Indian castes (see Cox 1948: 490). Castes were understood as clearly bounded, hierarchical and endogamous groups, with barriers restricting social contact – for example, eating together or even coming into physical contact. The castes were spatially segregated, for example, by railway tracks running through town: segregation was not absolute but was very marked. Black areas were poorer and less well serviced. Jim Crow laws enforced 'separate but equal accommodations' in jails, schools and public transport (see Fig. 7.1) – in reality far from equal, except perhaps in jail; informal but equally fierce segregation prevailed in washrooms, eateries and churches.

Sexual relations were an especially emotive domain, and policing of sexual boundaries was key to the whole caste system. As we have seen in previous chapters, race, sex and gender were closely related: racial difference was perpetuated through the regulation of sex and sexual reproduction, and this happened in gender-unequal ways. Interracial marriage was prohibited, by law. Sexual relations between black men and white women were particularly taboo, while relations between white men and black women were more common, most often transient and semi-covert, but sometimes involving long-standing relationships and children. Such offspring were automatically included in the lower caste. The categories of black and white were clearly bounded and defined by the 'one-drop rule', which had become law in several states after 1910.

Within each caste there were significant class differences: among whites, who formed about 25 per cent of the county population of Indianola or 50 per cent of Natchez town, there was a small elite or aristocratic planter class, a larger middle class and a substantial

Figure 7.1. At the bus station in Durham, North Carolina, 1940.
(Courtesy of the Library of Congress)

working class; among blacks, a large working class and a very small professional and land-owning middle class. In rural plantation areas relationships between whites and blacks tended to be those of landlord–tenant or master–servant; in urban areas relationships were more complex, as whites and blacks could sometimes work side by side. Still, caste membership was signalled by race and defined status: all blacks remained subordinate to all whites in some – ultimately symbolic – sense, even if their class status was higher.

Great emphasis was placed by these authors on a ritualistic racial etiquette, in which blacks had to constantly demonstrate deference to whites, by addressing them, for example, as 'mister' or 'sir', while whites would call blacks by their first names or 'boy' (for males) and would avoid using 'mister' if at all possible (e.g. by calling a middle-class black teacher 'professor'). Whites saw blacks as categorically inferior and essentially childlike. In this regard, elite and middle-class whites were depicted by the ethnographers as paternalist, while lower-class whites were described as virulently racist. Working-class blacks tended to be less antagonistic towards whites than middle-class blacks, who, although their lifestyle and values were closest to white models, strongly resented racial exclusion by the white middle class and certainly despised poor whites.

Maintaining segregation

How was this segregated and hierarchical system maintained, often in contexts in which black people were a majority? As with many hierarchical systems based on capitalism, the key mechanisms of exclusion were around economic ownership, education and political control. These excluded the white working class too, but they were skewed towards the exclusion of the black population. Ultimately, too, the system was policed by the threat and use of violence, in the form of threats, beatings – which included slaps and punches quickly and publicly delivered by whites to punish perceived 'sassiness' by blacks – and, in its most extreme form, lynching. Technically lynching was illegal and, as public criticism of the practice grew in the 1920s, threatened lynchings were quite often stopped by white citizens and law officers. But many went ahead, with the law turning a blind eye, so the threat was a powerful one. Mississippi had the highest rate of lynchings in the South, with some 500 black men (and 45 whites) lynched between 1882 and 1930, although the number of cases per year fell markedly after a peak in the 1890s (Tolnay and Beck 1995; Work 1931: 293). Even if a lynching did not take place, a black man accused of attacking a white woman could be tried and legally hanged within a matter of days.

In economic terms, many blacks in the 1930s cotton-belt were tenant farmers, who worked land owned by whites and depended on the landowners for housing, subsistence, seeds and tools. The tenant was given some of the proceeds of the sale of crops, once the costs incurred by the owner were deducted. This system was wide open to abuse by landowners, who routinely swindled their tenants. Tenants were white as well as black, but the racial etiquette and the ever-present threat of violence for 'not knowing one's place' as a black person skewed the system against black tenants, even if planters sometimes preferred black workers who were easier to bully. Tenants' only power was to move to a different landowner, but this was not easy when incomes were low and precarious.

A move to factory work was possible, but there black workers were allocated the lowliest jobs, and even then received lower wages than whites doing the same work; again employers sometimes preferred black workers as they could be disciplined and exploited more easily (Davis, Gardner and Gardner 1941: ch. 20). The organisation of labour into unions was strongly resisted by white employers as a threat to their dominance.

Politically, black voters were disenfranchised by a combination of a poll tax, which many poor people, including poor whites, could not afford, and by the selective use of a test in which the voter had to show an understanding of a passage in the US Constitution, to the satisfaction of the local electoral registrar, who used the test to exclude blacks. Southern politics had long been dominated by the Democratic Party, which had enshrined segregation into law: party officials controlled all nominations and made black voter registration very difficult.

In terms of education, Dollard (1937: 203) noted that the situation in 1930 was very different from that in 1870: in the United States as a whole the percentage of blacks aged five to twenty attending school had risen from 9% to 60%, although in Mississippi white school enrolment was at 82% compared to 60% for blacks. Greater education for black people was a major factor generating pressures for upward mobility. Yet he also observed that in Indianola black people received a training directed towards manual occupations, and had to make special efforts to go beyond this. The school system was entirely segregated, and Mississippi state expenditure per white pupil in 1930 was five times than for a black pupil; white pupils spent more days a year at school, while a white teacher taught almost half as many students as a black teacher and earned more than twice as much (Work 1931: 204–5).

The work of Dollard and Powdermaker drew on specific cotton-belt locations at a particular time, to create a picture of the US South that has proved influential, but which has been criticised on a number of grounds. The studies privileged race as the key divide of the South, overstating the rigidity of the divide, giving an impression of seamless white unity and especially understating the importance of class as a basis for conflict and alliance. While they acknowledged the class diversity of the whites, poor whites got little attention in these studies, coming across simply as virulent racists whose aggression was a psychological defence mechanism. But details of the studies themselves, and in particular the lesser-known study of Natchez by Davis and the Gardners, indicated that many blacks did not believe that poor whites were as much of an obstacle as middle-class whites. Meanwhile, poor whites were not necessarily so hostile to blacks, with whom they shared a class position and some religious practices. They actually felt intense hostility towards the white elite, especially as labour was squeezed by the economic crisis of the 1930s Great Depression and the beginnings of the mechanisation of cotton cultivation. In the face of this squeeze, any working-class unity between blacks and whites was not easy. The federal government stepped in with agricultural programmes, which gave favourable access to white farm workers. This helped to undermine interracial labour organisation, which existed for example in the shape of the Southern Tenant Farmers' Union (Adams and Gorton 2004).

Jim Crow segregation set the tone for the racial formation of the United States and was the context for the ensuing processes of black mobilisation, desegregation and racial

One result of these political changes was a paradox, already explored in Chapter 5. On the one hand, race as a concept remained central to, and overtly present in, US politics and society. On the other hand, as racial discrimination became increasingly unacceptable, race began to become increasingly awkward in, or even disappear from, some contexts of public discourse. The very mention of the term, or of racial categories, could imply that one was taking race into account, whether to discriminate against or in favour of someone else. Instead, it was more acceptable to talk in terms of culture, a discourse of difference that could include Koreans, Irish, Poles, West Indians, African Americans, Jews, etc.

Colour-blind practice, which evades mention of race or even power difference, fits into the basic principles of American liberal democracy and individualism, in which everyone is supposed to be equal and should be treated as an individual not as a member of a collective. Meanwhile, racial (and class) inequality have remained pervasive – or even increased in some respects. Racial identities, distinct from simply 'cultural' ones, also remain deeply rooted, marked by the particular discourse of natural–cultural difference, together with the use of the colonially derived categories of genealogical colour – black, white, brown – that, as I have argued in previous chapters, characterises race. But racial inequality and difference are both masked by colour-blindness, leading to a situation of 'racism without racists' (Bonilla-Silva 2003). This has to be understood not primarily in terms of people cynically denying being racist while covertly practising racism (although this can happen), but rather in terms of people self-consciously wanting to avoid racism, while also pursuing what they see as rational and proper ways to behave and defend their own interests, pursuits that can have racially discriminatory consequences, whether they intend these or not (Hartigan 2010: 11). The following sections examine processes of segregation in more detail.

7.4 Segregation in practice: 'the ghetto'

If the Jim Crow South is one key emblem of race in the United States, the other is 'the black ghetto', typically associated with the North – New York, Chicago, Detroit, Philadelphia – but which also exists in Los Angeles, Washington, Miami and Houston, among others. The black/white dissimilarity index for these major metropolitan areas was between 65 and 88 in 1990, although the figures had dropped to between 60 and 78 by 2010 (Population Studies Center 2010).

How do ghettos develop? In the United States the state has played a major role. In the early twentieth century racial zoning, which established white, black and mixed urban zones, was legal until 1917 and widely used in Southern cities. After World Wars I and II black migrants moved in large numbers into Northern cities, drawn by industrialisation. A variety of mechanisms, some with state backing, operated to create segregation within the low-income residential areas that most black incomers, poorly qualified, in low-paying jobs and suffering discrimination in the job market, could afford to live in (Fair Housing Center of Greater Boston n.d.; Massey and Denton 1993; Seitles 1996; US Housing Scholars et al. 2008).

- Someone selling a house would use a 'restrictive covenant' to oblige the buyer not to sell to a non-white family. These covenants were recommended by the Federal Housing Administration (FHA) and in use until the late 1940s.

- Black and other non-white incomers to a white neighbourhood were often subject to violence. This ranged from intimidation and harassment to house bombings (e.g. in early 1920s Chicago) and full-blown race riots (e.g. in Chicago in 1919 and Detroit in 1943).

- White individuals and real-estate agents could refuse to sell or rent to non-whites. This was made illegal in the Fair Housing Act (1968), but has continued in less overt forms. Real-estate agents, for example, still tend to 'steer' clients by race towards particular neighbourhoods, where they think the clients will be more 'at home'. Tests show that discrimination on grounds of race (but also nationality) still occurs in local housing markets and mortgage lending. In 50–60 per cent of cases in Boston African Americans and Latinos are shown fewer properties, asked more questions, 'steered' in some way, required to give substantial advance notice before viewings and offered higher loan rates and fewer discounts; even upper-income African Americans are eight times more likely to have high-cost loans than upper-income whites (Fair Housing Center of Greater Boston 2006).

- Neighbourhood improvement associations in majority white areas lobbied against public housing projects unless these were reserved for whites. Until such discrimination was outlawed after WWII, government housing authorities segregated such projects by race. As late as 1989 the Dallas housing administration was found to have deliberately, and illegally, maintained racial segregation through its public housing policy. Public housing projects and programmes offering housing assistance to poorer people are still often concentrated in minority neighbourhoods.

- From the 1930s the FHA instituted a 'redlining' policy, in which the creditworthiness of mortgage applicants was based, in part, on the racial composition of the neighbourhood they wanted to buy in. Black or mixed areas – ones that contained 'inharmonious racial groups' in the FHA's terms – were marked as high risk, with a 'red line' on the map, and mortgages there were more expensive or unavailable, restricting access to home ownership for black families. Between 1930 and 1950 60 per cent of homes purchased in the United States were financed by the FHA, yet less than 2 per cent of FHA loans were made to non-white home buyers. Black areas then suffered disinvestment and decay, making them unattractive to whites who could afford to live elsewhere and thus increasing segregation. Redlining by federally chartered savings institutions continued into the 1970s.

- Black families who tried to move into white neighbourhoods often faced obstacles (harassment, restrictive covenants), but sometimes were able to do so. In that case, white families tended to move out, precisely because the presence of black families in the block was likely to attract redlining or more generally because they feared that property values would fall. Real-estate agents capitalised on this fear with the practice of 'blockbusting', in which selling or renting some properties to non-white families on a block triggered an exodus of white owners. The estate agent could then buy cheaply and either sell at a good profit to incoming black buyers or more probably convert the civic associations, on the community board and in church. They shared civic rituals and festivals, whether these were nominally ethnic – such as the Colombian Independence Festival, the Korean Harvest Festival or the Black Heritage Book Fair – or non-ethnic, as in the Corona Heights Fourth of July Celebration, Memorial Day, Christmas Tree Lightings, Veterans Day, public school cultural events or Three Kings Day. People collectively struggled with problems around quality of life and urban services, and especially schools, education and youth. On the Community Board 'lines of race and ethnicity had become crossable', and in the area many whites had moved from 'categorical to personal relationships with individual African Americans and immigrants' because of their own experiences of interaction (Sanjek 1998: 330, 391).

Sanjek recognised that not everyone had changed in this way: the local real-estate industry, for example, was overwhelmingly dominated by whites, and the industry as a whole still worked in segregationist ways. A 1991 study in New York found that over 40 per cent of black buyers and renters faced discrimination (and Latin Americans faced even more), while the New York Public Housing Authority admitted as late as 1990 that it had been assigning tenants to housing projects on the basis of race, despite a federal warning in 1983 (Sanjek 1998: 150). Also, electoral politics were still shaped by race. Before 1965 electoral districts with big black or Puerto Rican populations were 'gerry-mandered' (had their boundaries altered) to dilute ethnic–racial voting blocs and keep black and Puerto Rican officials out of power. The 1965 Voting Rights Act banned this practice in order to increase minority representation, but 'redistricting' continued, although now with added lobbying by minority groups eager to maximise the chances of getting one of their representatives elected. But Sanjek found that, in 1991, incumbent officials, who were white, managed redistricting in ways that protected their own power and thus effectively undermined the non-white vote. In any event, it is clear that, in elections, people were assumed to vote, at least in part, along ethnic and racial lines. Overall, however, the message of Sanjek's study was one of gradual but effective desegregation. And, as racial segregation is a very powerful driver in the reproduction of racial inequality and racial difference, this indicated a positive trend in US society, one reflected in the gradual decreases in racial dissimilarity indices in most US cities over the last twenty years.

Black Corona

Sanjek admitted that his overall message was least true when it came to African Americans, who, in the area he studied, were only 4% of the total. They concentrated in a small area, centred on a housing project called Lefrak City, which was over 70% black in 1990. In the area Gregory studied, North Corona and East Elmhurst, just to the north, segregation was also more marked: by 1960 North Corona had become 50% black and in 1990 some census tracts in East Elmhurst had over 80% black population (Gregory 1999: 57, 111, 144). This, in itself, is highly indicative of powerful racial segregation, within the area of some four or five square miles covered by Sanjek and Gregory.

The processes that led to this situation are familiar: blacks migrated into neighbourhoods they could afford and they remained a minority before WWII. Although most were poor – 77% of black women in Queens worked as domestic servants, compared to 7% of white women – a black middle class of white-collar workers began to emerge early on, with their own clubs and newspapers. After 1945 immigration increased and so did segregation: white flight meant that 70% of Corona's white population left during the 1950s, moving to other suburbs; Corona shifted from being 26% black in 1950 to 50% black in 1960; the area was shaped by the combination of forces outlined above: redlining, block-busting, steering, etc. 'The mere presence of whites in significant numbers assured better public services', so the white exodus meant that conditions deteriorated as the tax base declined and businesses closed down (Gregory 1999: 57). Schools became more segregated as white pupils left the area.

The Lefrak City housing complex was a very clear example of emerging segregation. Built by property developer Samuel Lefrak in 1964, it was in 1970 about 70 per cent white. In 1972 Lefrak was accused by the government of discriminating against black renters in New York. He agreed to end practices that could be used to racially discriminate – such as not publicly posting all vacancies, not processing applications on a first-come-first-served basis, not applying the same credit checks to white and non-white applicants, and not spending the same on the maintenance of majority-black and majority-white buildings (Sanjek 1998: 55) – and he agreed to help black families move into his buildings. By 1976 blacks made up nearly 80 per cent of Lefrak City tenants.

White tenants fled, but not explicitly because of race, as such: as one said, 'I keep saying to you that it had nothing to do with colour. There were always black people here when I moved in. They were friendly … high-class people … But when you saw these people coming in – in their *undershirts*, and their *hair*, and their *staggering*. It really was the most *horrible* thing you ever saw' (Gregory 1999: 113, emphasis in the original). This was apparently a discourse about class and culture, but it was clearly also about race, as ideas about colour, class and culture merged together. Lefrak City rapidly became labelled as a 'welfare haven', a black ghetto enclave that threatened its neighbours with drugs, crime and social disorder. In fact, in 1976 only 3 per cent of the tenants were on welfare, but in the imagination of many whites race, poverty and social pathology were intimately linked, as a natural–cultural hybrid, and they left in large numbers, feeling that the city was using rent subsidy schemes to dump problem black tenants on the apartment blocks.

Gregory is careful to avoid, however, the image of a homogeneous black ghetto. He is very conscious of the stereotypes of the dysfunctional ghetto and is aware that, even though anthropologists had emphasised that 'ghetto culture' and poor black family structures were a resilient and adaptive response to structural deprivation and racism, the very notion of a ghetto culture tended to set these areas apart and homogenise them (Gregory 1999: 9; Kelley 1997). Gregory shows that the black population of Corona was far from homogeneous, as there had long been a black middle class, which had expanded post-Civil Rights. Nor was it enough to think in terms of a clear split between the poor ghetto dwellers and the black middle class who fled the inner city. In Corona black people of different class status lived in the same area and were linked by cross-cutting ties of community activism and cultural practices, with middle-class blacks taking a lead in organising to improve quality of life.

For whites in this area, whiteness is complicated into a 'hundred shades of white', in part because 'hillbillies', while white, were also originally from the South and seen, stereotypically, by other whites as dirty, 'low-down' and disreputable – characteristics typically attributed to black people. 'Hillbillies' were seen as not properly white, in a cultural sense. This ambiguity was found among the hillbillies themselves: as one man said, 'you don't have to be black to be a nigger. Niggers come in all colours. We are all coloured' (1999: 116). This was mainly a use of the term 'nigger' to label 'bad behaviour': as we saw in the Andes (Chapter 5), if you act like a black or a white, you become closer to that identity. But it also indicated a porousness of racial boundaries. In one joking exchange between Bill, a black man, and Donnie, the latter first asserted he was 'an Irish hillbilly' and then said: 'I believe I've even got some nigger in me, too.' He backed his claim – met with laughter by Bill – with the fact that he liked Southern food. When he mentioned chitlins (pig intestines), Bill replied, 'Chitlins too! Well then I believe you are kin of mine' (1999: 114).

This was all a joke, but it was part and parcel of a certain cross-racial intimacy in Briggs, where whites had little to defend in terms of territory, property, propriety and middle-class moral values. Racial identities and discourse had an overtness to them – as in the frequent use of 'nigger', which in many contexts has become a taboo word – but they were rarely used to characterise collective categories of people in a stereotypical way. Rather, they were always available as a mode of interpretation of specific daily events – a joke, some music, a particular bit of behaviour, etc. For example, Jerry, a white man, accused his brother of being 'a nigger' when he thought he'd stolen a music cassette from him; and called another white friend 'nigger' when he played guitar well. On both occasions this was in the presence of a black friend, who remained unmoved. Or racial language might be used to interpret more violent events, such as a mugging. On the other hand, people might reject racial language as inappropriate to interpret some situations. For example, when someone was causing trouble in a bar and hurling racial insults (including 'nigger') he was seen as just causing trouble, not expressing some racial tension; his use of racial language was seen as just an easy attempt to rile people up. Thus race was not about fixed, absolute identities, but about a malleable language of reference, a semantic domain that could be used selectively, depending on the context.

Hartigan's point is that the use of 'nigger' is more complex than 'it's just racist', even when used by whites in Briggs. The term always connotes a broad structural sense of power and hierarchy – a polarised difference between black and white in US society – and can be a direct racial insult, but it is not always reducible to that when deployed by a specific person in a particular context. It also reflects the fact that whites in Briggs were seen as, and saw themselves as, racially ambiguous, because they lived in a semi-derelict poor and racially mixed neighbourhood (1999: 112–28). Briggs indicates the complexity of racial segregation and identity in the inner-city United States: we can see elements of the cross-racial conviviality that Sanjek found in racially mixed Elmhurst and we can see some tendency for racial identities to lose a little of their typical US clarity and boundedness. But overall racial hierarchy continues to structure local interactions in a powerful way.

7.5 Latinos and brownness

Under conditions of strict segregation in the United States, racial identities were clearly bounded – although never completely so, as some 'black' people could always 'pass for white', even if they had 'one drop of black blood'. The example of Briggs, Detroit, shows that, in the particular context of a post-war, racially mixed, inner-city neighbourhood, the performative dimensions of racial identities – more clearly visible in some Latin American contexts – can become visible. The natural–cultural construction of race comes into clearer view as behaviour racially darkens whiteness. This possibility could, in theory, become more prominent with the increasing presence of the Hispanics and Latinos – to use the generic labels created to describe them – who have, since 1965, become a major feature of the US racial panorama and who are, in some sense, racially intermediate between black, white and American Indian (Hartigan 2010: ch. 6). (For reasons of space, I will not explore Asian immigrants and Asian Americans in this book.)

LATINOS: DEMOGRAPHIC AND ECONOMIC POSITION

- Hispanics, the census category coined in 1980, now constitute over 15% of the US population; 65% of Hispanics are of Mexican background and 9% of Puerto Rican background.
- In the 2010 census 53% of people of Hispanic or Latino origin identified as white, 37% as 'some other race', 2.5% as black, 1% as American Indian and 6% as two or more races (of the latter the majority identified as being a mixture of white and 'some other race') (Humes, Jones and Ramirez 2010).
- In general, Hispanics are less segregated than blacks (although slightly more than Asians), with dissimilarity indices below 62 in the major metropolitan areas in 2010 (Population Studies Center 2010).
- Economically, Hispanics sit slightly above the black population on various measures – poverty rates, median household income, unemployment – but not by much (National Urban League 2014). Both populations are clearly below both whites and Asians.

Despite the fact that a majority of Latinos identify as white in the census, almost 40 per cent refuse to classify themselves in the main racial census categories and instead tick the 'some other race' box (Rodríguez 2000). Latinos as a category constitute an ambiguous presence for the dominant racial binaries of the United States. After the United States annexed parts of northern Mexico in 1848, Mexicans in the area were promised citizenship and legal whiteness; facing de facto discrimination, they fought for their legal whiteness to be recognised. From 1940 to 1980 the census classified Mexicans and other Latin Americans as white. However, outside the legal fiction of whiteness and the census, Mexicans in the South and South-west, especially those with darker skins, found themselves subject to Jim Crow-type segregation. If they married a white person they could become entangled in laws against interracial marriage. At the same time Mexicans especially were being cast as 'illegal aliens', a term fraught with racial connotations of non-whiteness (Menchaca 2007; Stephen 2007: 222).

One study shows that Mexican Americans who identify as white on the census form do so as part of a defensive, race-evasive strategy to downplay the persistent racism they experience at the hands of others; they do not usually identify as white in other contexts, they speak Spanish and are proud of their Mexican heritage. Identifying as white is not evidence of a simple process of assimilation; these people are invoking the earlier, hard-won legal definition of Mexican Americans as white citizens: as one man put it, 'I'm white, cause I'm an American, right?' Their race evasion is born of pain, not privilege. Mexican Americans who identify as 'other' tend to be more assertive about the racism they experience, resisting it more openly with an affirmation of Mexican identity; they align themselves with Mexican immigrants and even African Americans (Dowling 2014). This assertive stance is reflected in the fact that, from the 1960s, some Chicanos – a term designating people of Mexican descent born in the United States, particularly in the western states – began to mobilise around Brown Power, as a counterpoint to Black Power. Pride in brownness has been taken up again in the 2000s, as a positive image for the future of the United States rather than an ethnic movement (Rodriguez 2002).

It is true that the black–white binary has a marked impact on Latinos. Various studies have shown that white Latinos, especially US born, marry non-Latinos (mainly white) more than non-white Latinos; that darker-skinned Latinos have poorer educational and income-earning outcomes than lighter-skinned ones; and that lighter skin tone acts as 'social capital' which aids Latino women's upward mobility (Wade 2009b: 233). Ethnographic evidence also indicates that darker Puerto Ricans are surprised to find themselves classed as black in mainland US terms, but often end up more or less acknowledging the force of that classification, even if they retain a strong Puerto Rican identity that distinguishes them from other black Americans (Rodríguez 1994).

Yet Latino remains a resilient and growing category that cannot be organised by standard US racial categories. Although Hispanics can be 'of any race' in census terms, and although a majority are white, there is a sense in which Latino or Mexican becomes a kind of intermediate racial category, in between black and white.

Dominican women and racialised aesthetics

For example, Dominicans in the United States rarely identify as black, in keeping with their preferences in the Dominican Republic too, where blackness is associated with neighbouring Haiti, with which the Dominican Republic has had an antagonistic relationship. Candelario explores how Dominican women in New York cultivate a whitened body aesthetic in beauty salons, especially in terms of achieving *pelo bueno* ('good' or straight hair). Dominant gender ideologies that value women's personal appearance interact with racial ideologies that value whiteness, to create a specific concern with skin colour and hair texture, especially among women who have the economic means to afford beauty treatments (Hunter 2002; Moreno Figueroa 2012). Yet these Dominican women did not want to look too white: *pelo bueno* was also *pelo muerto* (dead hair). The best look was one that was, in their eyes, 'Hispanic' or 'Dominican', and this was understood to be 'a mixture of black and white': skin that was light but not too white; hair that was straight, but not lifeless and limp; lips that were not thick, but not too thin either. This was having a bit of 'black behind the ears' (Candelario 2007: 230). Hispanic operated as a 'fluid middle' between black and white, which allowed them to avoid blackness, seen as low status both in the United States and in the Dominican Republic, but also not to capitulate to an Anglo-Saxon form of whiteness associated with US imperialism and racism. In that middle ground women would also adjust their appearance to suit specific contexts: straightening their hair more for fancy occasions and letting it curl more naturally for informal events, sports, recreation and so on.

The Puerto Rican 'nation' in Chicago

Puerto Ricans in Chicago are also shaped above all by a sense of their own national identity, which is certainly influenced by whiteness and blackness, but does not fit neatly into two opposed categories. The neighbourhoods (*barrios*) of Humboldt Park, Logan Square and West Town studied by Ramos-Zayas (2003) became Puerto Rican strongholds in the 1940s. By the late 1990s locals there felt under threat from various groups of incomers: (a) African Americans were black, which was an important symbol for some Puerto Rican nationalists in the area, but American blackness was seen as more exclusive and did not respond to a Puerto Rican history of race mixture; black Americans were also associated with a consumerist capitalism that it was feared might undermine Puerto Rican traditions; (b) whites were seen locally as inauthentic Bohemians who had money but lacked respect and didn't wash enough; (c) other Latinos, mainly Mexicans, were often thought to be illegals, unlike Puerto Ricans. Local Puerto Ricans also had differences with island Puerto Ricans, whom they saw as an overly whitened and colonised elite, which assumed ownership of a Puerto Rican culture that they represented with consumerist and inauthentic symbols that did not evoke Puerto Rico's history of anti-colonial and working-class struggle. The *barrio* nationalists combined a very particular combination of class, anti-colonial and racial discourses to enact a nationalism they saw as authentic; at the same time they valued hard work and the possibility of success and progress – nothing less than the American Dream – as a way of contesting images of Latinos as lazy welfare parasites. Blackness and whiteness were important factors in constructing a Puerto Rican identity that could not be encompassed by either or both.

Mexican workers in a Chicago factory

The complex way Latino racialised identities operate in the US environment is shown nicely by De Genova's study of Mexican workers in the Chicago factories where he taught English classes. De Genova describes a series of joking encounters around the issue of race, all connected to a Mexican man, Gonzalo, who deployed an ironic, subversive style of humour (De Genova 2005: 174–85):

- Gonzalo joked with a US white worker that only he, Gonzalo, could go to the language classes because he was white – implying that the US man was not: 'school is only for white people, no *negritos* – you can't go'. Of course, school was actually for Latinos, who might very well not be fully white by US standards, but it also meant a temporary respite from work, a privileged leisure associated with whiteness.

- Gonzalo called his best Mexican friend, Osvaldo, his *negro*, not because he was particularly dark skinned, but because he worked 'like a slave', supplying machine operators like Gonzalo with their materials.
- On learning where De Genova lived on Chicago's South Side, Gonzalo said, 'I don't like the South Side, because there's a lot of niggers on the South Side.' However, Gonzalo was perfectly aware that the specific area that De Genova had named was an entirely Mexican neighbourhood. He seemed to imply that all Mexicans were like blacks.
- Osvaldo had grown up in a rural town where he had picked up phrases in a local indigenous language, Nauhuatl, which he would enunciate on Gonzalo's demand, producing much laughter from his companions: the comic use of indigenous words enacted a typically jocular and condescending anti-indigenous Mexican racism.
- Gonzalo referred to his fellow workers as 'niggers', although there were no African Americans there. He even said that his boss pushed him hard, because he was a darker colour – 'because I'm a nigger'.

These Mexican migrants jokingly used a language of blackness, whiteness and indigeneity to locate themselves in highly ambiguous ways in the US (and Mexican) racial panorama. Overall, they distanced themselves from American blackness (and Mexican indigeneity), yet they also continuously racialised themselves. They, or Gonzalo at least, might claim whiteness – perhaps in the process ironically undermining a white worker's racial identity – while also recognising at times that they might be seen as black or like blacks, because they had to work 'like slaves'. Working hard, and particularly being obliged to do so by the bosses, was labelled 'slavery' and always linked to blackness. White workers also had to work hard, but, for all Gonzalo's ironic subterfuges, they shared with the white bosses a racial identity that was relatively solid.

The opposite of working hard was being lazy (*huevón*) and this was both bad (in contrast, Mexicans were hard workers and would get ahead) and good (someone who could be lazy, *for a while*, was resisting the bosses' impositions and standing up for himself as a proper Mexican man should). Laziness was thus ambivalent, and also had ambivalent racial meanings for the Mexicans: to them the white bosses were lazy, because they worked in the office and ordered other people around; the African American workers were lazy, because they didn't knuckle down and work hard to get ahead, and instead depended on government welfare and employment (black 'laziness' was usually labelled as being *flojo* – weak, flaccid – rather than *huevón*, which had connotations of masculine autonomy) (De Genova 2005: 189–95).

Gonzalo's subversive jokes and the racialised ambivalence of work and laziness could play with racial identities in apparently contradictory ways because, in the end, 'Mexican' was racially intermediate between black and white, but powerfully structured by these poles. Mexican identity was also inflected by the category of Native American, but less in relation to the US version of that figure and more in relation to the Mexican *indígena* or, more pejoratively, *indio*. Stephen's study of indigenous Mexican migrants to the United States clearly shows that people identified by other Mexican migrants as indigenous – because they come from Oaxaca (a Mexican region with a large indigenous population), because they 'look indigenous' (they are short and dark skinned), or because they sometimes speak an indigenous language, such as Mixtec or Zapotec (usually in private) – are subject to disparaging comments at work and in schools. In one agricultural labour camp the 'Oaxaquitos' (the little people from Oaxaca) lived in separate cabins where, according to the others, 'they don't even speak Spanish' (Stephen 2007: 216). This discrimination has helped to give many Mexican indigenous people a stronger sense of their own identities, and this is reflected in the creation of numerous organisations, whether ethnically based or broadly based, such as the Front of Binational Indigenous Organizations (see also Fox and Rivera-Salgado 2004; Kearney 2000).

Latino identity and US racial binaries

Latino or Hispanic identity cannot be reduced to clear US racial binaries – partly because both terms encompass a huge variety in national, ethnic and racial terms and, as labels, do not necessarily command much adherence from the varied people they attempt to encompass – but they are deeply shaped by these binaries, not only in the US context, but also in Latin America, in the sense that mestizo identity always draws its meaning from its originary components of black, white and indigenous (Dowling 2014). Thus, 'the "Latino" (or "Hispanic") label tends to be always-already saturated with racialised difference' (De Genova and Ramos-Zayas 2003: 2).

The growing presence of a highly diverse 'brown' category, which is not reducible to black, white and American Indian and which is less physically segregated, clearly complicates the classic US racial binaries. In US high schools in California, for example, some young white people dress and act 'gangsta' in ways that identify them with their Latino peers, at least superficially. This creates porous boundaries between white, brown and even black. In Indiana a group of white US girls triggered an 'imitation race war at their virtually all white high school [in 1994] by dressing "black"' (Bernstein 2003). But US racial binaries still work to organise and structure this new brownness. The fact that, economically, Latinos are clearly much nearer to blacks than to whites also weakens the desegregation effect that Latinos might have on US society.

7.6 South Africa

My purpose in moving the focus to South Africa is not to give a thumbnail overview of a different context, but to explore a specific comparative theme. In significant ways the United States and South Africa are similar in the sense of having established legally underwritten systems of white dominance and racial segregation and then undergone a process in which these were dismantled and attempts made to reverse their legacies. In the case of South Africa the difference that interests us is that the change, which took place with the fall of apartheid between 1991 and 1994, involved a radical change of economic policies in ways that have actually increased overall economic inequalities in the country. The upshot has been that, while inequality between racial – now called population – groups has decreased slightly, inequality within racial groups, and especially among black Africans, has increased markedly. That is, the black middle class – which existed under apartheid, but was severely constrained (Kuper 1965) – has expanded and become wealthier, while most Africans remain in poverty.

RACIAL INEQUALITY AFTER APARTHEID

- In the 2001 and 2011 censuses Africans are 80% of the total (up from 70% in 1970), with whites at 9% (down from 17%), Coloureds at 9%, and Asians and Others at 3%.
- African per capita income was 7–9% of that of whites from 1917 until 1993; between 1993 and 2000 it varied between 11 and 16%; and in 2008 it was 13%.
- Between 1996 and 2001 Africans and Coloureds continued to form over 97% of the bottom 50% of per capita earners. However, Africans made some gains in the upper-income brackets in relation to whites, reflecting the growing African middle class and elite.
- Africans account for over 95% of those in poverty in 1996 and 2001, with whites and Asians making up less than 1% together and Coloureds accounting for the remainder; between these years poverty increased for Africans and Coloureds, but stayed the same for whites. By 2008 Africans were still 93% of those in poverty (Leibbrandt et al. 2010; Leibbrandt, Woolard and Woolard 2009; Seekings and Nattrass 2008).
- In 1992 the management-level occupational categories were 93% white and 2% African; by 2000 the figures were 80% and 10% (mostly men in all cases) (Moleke 2003).
- Largely due to economic inequality, residential segregation has remained very high, with dissimilarity indices for all racial groups in urban areas dropping a little between 1991 and 1996, but still 94 for whites, 87 for Africans and 80 for Coloureds. Cities such as Johannesburg, where Africans moved into the city centre, showed patterns of white flight to gated suburban communities (Christopher 2001; Seekings 2010).

With the growth of an African middle class and elite it is clear that race and class are less closely linked than they used to be. The overt discrimination of the apartheid era no longer accounts in the same way for racial inequality. This does not mean that racial discrimination has disappeared: after all, whites still control a huge slice of the economic opportunities in the marketplace and it seems unlikely that their prejudices would have vanished in twenty years (see Fig. 7.2); urban white flight suggests otherwise, as does the continued significance of racial difference in public life (see below). But, as in the United States and Latin America, a huge role is played by the fact that most black South Africans are poor and live in precarious conditions. There continues to be a very powerful structural link between poverty and blackness, which race-based affirmative action programmes only change at the margins, generating some wealth for a minority of Africans, while business as usual in the wider economy exacerbates inequality and poverty overall, affecting a tiny minority of whites as well – whose position nevertheless attracts media attention – but with the main impact affecting African and Coloured people. This structural link perpetuates black disadvantage by feeding racial stereotypes and reinforcing

Figure 7.2. Two weeks after the city allowed blacks to travel on 'Whites Only' buses, blacks and whites wait for the same bus.
(Joanne Rathe/The Boston Globe via Getty Images)

the social distances and barriers that impede black upward mobility. Changing that link means both (a) equalising the distribution of racial groups across the class structure and (b) reducing poverty and inequality in the society overall, as even if 80 per cent of the people in poverty were African and 10 per cent of them white, in accordance with their demographic weight, this would still maintain a problematic overlap between blackness and poverty.

After apartheid: identity and difference

Turning to racial identity and difference, South Africa post-1994 shows a brief period in which the significance of racial difference became more muted. Under Nelson Mandela the ANC took a tolerant stance, emphasising national unity and reconciliation; a 'sunset clause' allowed white government officials to remain in post for five years; Mandela donned the rugby shirt of the famously nearly all-white Springbok team at the 1995 Rugby World Cup. The Truth and Reconciliation Commission (1996–8) tended to avoid politicising race, in part because if racism were recognised as a legitimate political stance, it could be the basis for amnesty; thus racist murderers could be amnestied as political militants. But in the process of demoting racism to individual prejudices and motives, the reality of South African racism as a political system was masked (Wilson 2001).

By 2000, with Mandela and others getting increasingly restive with what they saw as white recalcitrance, a process of re-racialisation of the public sphere was beginning. White because ideas of racial difference and racism remained and have resurfaced in the context of increasing privatisation and commercialisation of the economy (Clealand 2013; De la Fuente 2001). Racial inequality in Latin America, the United States and South Africa is also reproduced by specifically racialised mechanisms, never independent of class mechanisms, but not reducible to them. This is evident in the way racial difference, often phrased as cultural difference, structures public debates about politics (do affirmative action and indigenous mobilisation entail reverse racism? Should we adopt colour-blind policies and language?) and everyday calculations about how to live life – for example, in a university residence.

The similarities between the three places are striking and speak to both common roots in colonial oppression and liberal political principles, and to the globalisation of race, anti-racism and multiculturalism. But the differences must not be overlooked, as race is always place-specific – which means attending to differences not just between these three places, but within them too. In simplified terms we can say that Latin America has been characterised by a history of mixture, while the United States and South Africa have been marked by a history of segregation. The latter two are distinguished primarily in terms of the majority native status of South African blacks, which has led recently to their political dominance, in contrast to the minority status of US blacks, the continued political and demographic dominance of whites, and the growing importance of a 'brown' middle. This is a useful shorthand, but it hides a wealth of variation within each area, of which this and the previous chapter have given just a hint.

FURTHER ACTIVITIES

1. Explore the 1930s ethnographies of the Deep South mentioned in this chapter to get a more detailed sense of how race was lived in that context. Classic novels about the period, such as Harper Lee's *To Kill a Mockingbird* (about the Deep South), or Richard Wright's *Native Son* (about Chicago), may also be useful. Classic films, such as *In the Heat of the Night* (about the Deep South in the 1960s), are worth a viewing.

2. Pursue some of the references listed in the section on 'Segregation in practice' to find out more detail about how racial segregation worked and continues to work in the housing market.

3. City-Data.com (www.city-data.com/) gives access to very detailed information about the United States (racial demographics, unemployment, sexual diversity, etc.) at levels ranging from the entire country to individual census tracts. You can explore the current profile, including the degree of racial segregation, of some of the US neighbourhoods mentioned in this chapter.

4. Explore the debate about South African anthropology sparked by Francis Nyamnjohm's (2012) article. For example, look at the replies in *Africa Spectrum* 48(1) 2013. You could compare this to the 1995 furore around Deputy Vice Chancellor William Makgoba at Witwatersrand University (Wits). Look at the *Southern Africa Report* 13(2) 1998.

8 Race in Europe: immigration and nation

If a key frame for understanding the reproduction of racial difference and racial inequality in Latin America was the idea of mixture, while for the United States and South Africa it was segregation and desegregation, then race in Europe is overshadowed by the theme of immigration. This does not mean that that immigration has been unimportant in Latin America or the United States – or that mixture and (de)segregation are not significant in Europe – but immigration has provided the dominant frame for recent ideas about race in Europe in a way that exceeds both other regions (Silverstein 2005). The idea that racialised minorities are, in one way or another, related to the image of 'outsider', and that the nations of Europe are, in a deep historical sense, 'white', has profound roots in the region. This is despite the fact that many countries have been linked into centuries-old transnational networks of colonialism and empire, which brought non-whites into their territories from an early date, and despite the fact that post-WWII immigrant communities have now given rise to third-generation offspring who are, in most instances, national citizens of the countries their grandparents migrated to. This said, to grasp race in Europe we need to extend our gaze beyond post-WWII immigration. In fact, ideas about race in Europe draw heavily on other, linked, histories, which create the context from which more recent immigrations take their meaning.

8.1 European histories of race

First, Western Europe has been the source of the colonial and imperial enterprises that were constitutive of modern racial formations. European thinkers and scientists were at the centre of the production of knowledge about human diversity and race, drawing on information derived above all from European colonies (see Chapters 2 and 3). Europe provided the colonists who conquered and settled the Americas as well as the colonial staff who administered and governed colonial possessions, but colonial Europeans and their colonially born descendants also circulated through Europe, and colonial soldiers, traders and administrators frequently returned there. Colonially produced commodities, such as sugar and tea, of course circulated through Europe and globally, but non-white people from the colonies also went to Europe, whether as visitors, ambassadors, servants, workers or treated as zoological specimens. These movements created dense networks of material and intellectual exchange and brought colony and empire into the heart of

Europe, even while the plantations, slaves, indentured labourers, rebellions and repressions were located mainly outside Europe (Gilroy 1993; Linebaugh and Rediker 2000; Matory 1999; McClintock 1995; Stoler 1995; Weaver 2014).

This colonial frame, conceptually dividing white Europeans from their colonised others, has been vital for constituting the post-colonial immigrant to European metropolises as a racialised outsider. As we will see, it is also a history that is routinely obscured, such that the non-white immigrant's presence in the metropolis is rendered suspect, rather than being tied to these colonial connections.

Second, there is an internal history to race in Europe, by no means disconnected from the colonial history (see Chapters 2 and 3). This begins with medieval ideas about the cleanliness of blood, the 'taint' of Jewish and Moorish ancestry (or *raza*), the constant medieval and early modern definition of Christian Europe in (often hostile) opposition to Muslim others to the east and south (Muslims who would later become colonised others), the seventeenth-century 'race wars', and the English classification of the colonised Irish as an inferior race. It extends, in the nineteenth and early twentieth centuries, into scientific classifications of the different 'races of Europe'; into virulent anti-Semitism, with Jews now seen as a biological race; and into programmes of eugenic reform, which construed the lower classes as inferior in quasi-racial terms and potentially damaging to the racial stock of the nation. This internal history is often seen as culminating in the Holocaust, as the ultimate expression of a war against 'the enemy within', who threatened the racial integrity of the nation. Part of the aftermath of the Holocaust has been the well-known tendency to remove race from public discourse, or to name it mainly in relation to anti-racism (Goldberg 2008: 156).

The inward-looking history of race was, in some ways, submerged in the context of post-WWII, post-colonial immigration: the migrants were seen by Europeans as coming from and belonging to the outside, and opposed to white natives as an undisturbed whole. But the internal history of race interweaves closely with the post-colonial panorama. For example, right-wing, neo-fascist groups that target immigrants and non-whites are often also anti-Semitic. The figure of the 'Muslim', who was important for defining medieval Christendom, later became a colonised outsider and, from there, has recently re-emerged as a perceived internal threat – whether to European national security (after the 9/11 attacks of 2001 and the London bombings of July 2005), European native cultures or tolerant liberal values. The threat is now seen to emanate from home-grown Muslim radicals, who are national citizens, as much as from immigrants. Meanwhile, the idea of white natives as a homogeneous category is fractured by perceptions of class difference, which, in recent versions, often cast the white working class as liable to be more racist and less tolerant than the middle classes, thus damaging a comfortable multiculturalism and rendering the white working class as a kind of disaffected 'enemy within'.

8.2 Issues in post-colonial migration in Europe

European countries have been a destination for people from Asia, Africa and the Americas for centuries: for example, Asian and black sailors formed small communities in British port cities from the eighteenth century; Vietnamese students were a presence in Paris in the nineteenth century. Substantial immigration occurred after WWII, in the wake of decolonisation, when a number of European countries also experienced labour shortages – at least until the economic downturn of 1973. Immigration has thus been integrally linked to the changing economies of European countries, where there was demand for cheap labour, supplied by people perceived by employers as newcomers amenable to work discipline and flexible labour regimes (Cohen 1987, 2007). In some cases immigrants did work that white native workers were reluctant to do. By 2000 European countries with significant percentages of migrants from Asia, Africa, Latin America and the Caribbean included those enumerated in Table 8.1.

Immigration has created an overall tension for European majority populations between rejection and an often grudging acceptance. Policy in much of Europe has been a balance between (a) trying to meet labour needs while also controlling immigration, especially from ex-colonial regions; and (b) trying to integrate the immigrant population and especially their locally born offspring who are often national citizens, through varying versions and combinations of multiculturalist, anti-racist and assimilationist approaches. On the other hand, minority groups have been keen both to claim rights

TABLE 8.1 Foreign-born population aged fifteen and over by country of residence and region of origin, in thousands, c. 2000

	Africa (Af)	Asia (As)	Latin America and Caribbean (LAC)	Total AsAfLAC	Total pop.	AsAfLAC as % of total	Foreign-born as % of total
France	2745	423	85	3253	48068	6.8	11.7
Netherlands	214	323	291	828	12733	6.5	11.2
UK	763	1475	324	2562	47684	5.4	9.4
Sweden	57	225	56	338	6463	5.2	11.4
Portugal	332	16	67	415	8700	4.8	6.7
Russia	13	5168	8	5189	119893	4.3	8.9
Belgium	232	62	20	314	8492	3.7	12.0
Norway	29	93	14	136	3667	3.7	8.3
Spain	372	79	725	1176	34848	3.4	5.5
Denmark	26	93	8	127	4359	2.9	7.4
Germany	178	966	53	1197	68113	1.8	12.5
Italy	408	189	220	817	48892	1.7	4.1
Greece	51	84	6	141	9273	1.5	10.8
Austria	22	59	10	91	6679	1.4	13.8

Source: Dumont, Spielvogel and Widmaier (2010).
Note: In this source Turkey is a European country and not included in Asia; at this date Germany had about 1,189,000 Turkish-born residents (OECD n.d.).

and integrate culturally, and to build resilient identities and forge new cultural horizons and social realities for themselves and the societies they live in. The latter path, in particular, effectively rejects the key conceptual opposition between integration and non-integration by positing a social and cultural re-engineering of the society in which the minorities live. This transformation involves both reimagining the nation within its boundaries and rethinking it as a bounded unit, due to the diasporic and transnational connections that immigrant communities bring with them. Several key areas serve as contexts for grasping the complexities of this tension: for examples I will focus on Britain and France, as both have large immigrant populations, but have different policy approaches to them.

Immigration and control

As outlined in Chapter 5, while post-WWII immigration policies have tended to avoid explicitly racial exclusion – with the United States opening up its borders to international migration after 1965 and receiving many millions of Asians and Latin Americans – in Europe countries such as Britain and France began to brake immigration in the 1960s and 1970s, using superficially non-racial criteria, and to deport more illegal immigrants in the ensuing decades. For many Western countries the idea that immigration must be restricted is taken for granted, and an open-borders immigration policy attracts little support (Carens 2003). In both Britain and France immigration policy attempted to minimise entry 'of colonial peoples that policymakers believed to be too different and difficult to integrate' (Schain 2012: 144). Such measures limited the expansion of the immigrant population in the short term, but the longer-term trends have been rather different in each country: in France the foreign-born population as a proportion of the total has decreased slightly between about 1982 (6.3%) and 2010 (5.7%), with North Africans also decreasing proportionally (from 2.5% to 1.7%); in Britain the foreign-born population has increased (from 6.2% in 1982 to nearly 12% in 2010), with immigrants from South Asia, Africa and the Caribbean increasing from 0.5% of the national population in 1982 to 3.3% in 2009 (calculations based on Schain 2012: 68, 145).

While Britain has allowed its immigrant population to grow substantially, overall in Europe there is a growing focus on control and surveillance, sometimes summed up in the image of a 'fortress Europe', which is more impermeable to non-white than to white people, without race being an official criterion of exclusion. For example, the European Union's Black/White list (later renamed Positive/Negative) specified the countries whose citizens need and do not need a visa to enter the EU: although the Black list includes Russia, while the White list includes Barbados and Panama, overall 'it is striking how often the two list systems will have the effect of distinguishing between white and nonwhite entrants to the EU'. Alongside these lists, EU-wide databases, such as the Schengen Information System and the Visa Information System, amass biometric and other data on people (and objects) crossing borders, with a view to controlling illegal migration and other crimes. Although race is officially absent in these scenarios, the data allow a practical focus on 'visibly different' minorities (whose visible difference may be constituted by cultural traits, such as beards and clothes, as well as skin colour) and allow the movements of people to be tracked, and the data shared by European states and used by EU border security agencies such as Frontex (M'charek, Schramm and Skinner 2014b: 475, 477).

Citizenship, anti-racism and multiculturalism

Countries have varied in terms of how easily immigrants and their children can become legal citizens: Germany and Switzerland have tended towards seeing immigrants as temporary sojourners; Britain and France have generally been more open. Overall, avenues to citizenship always exist and have given rise in many European countries to substantial minorities of non-white citizens, who are mainly the descendants of their immigrant parents. Alongside this, there has been a recent tendency in many countries – including France, which adhered to an open, assimilationist, Republican model – to constrain access to citizenship by imposing complex requirements, tests and interviews that the applicant pays for and that require the applicant to demonstrate 'integration' (e.g. through knowledge of the official language, the national culture, having a stable job, etc.) or even 'loyalty'. Various countries – such as Britain in the British Nationality Act (1981) – also put restrictions on gaining citizenship by *jus soli* (right by virtue of birth on national soil) (see Chapter 5). Race and ethnicity are not mentioned here, of course, but the effect is to constrain access to citizenship, especially for poorer and less educated immigrants who tend to be non-white, and to construct a version of the country based on a traditional, national culture, perceived as white – or at least not 'coloured'. Limiting immigration goes hand in hand with limiting access to citizenship.

Meanwhile, in line with global trends, countries adopt some version of anti-racist legislation. Some countries, such as France, may focus on outlawing racist expressions (including Holocaust denial in 1990 French legislation), although this always runs up against the right to freedom of speech (Bleich 2011). Others focus more on discrimination in the markets (jobs, housing, education, etc.) on the basis of criteria such as race, colour, origin and often nationality. Britain was explicit in outlawing racial discrimination by introducing the Race Relations Act in 1965; the 1976 version of the law went further by banning 'indirect discrimination', in which, for example, an employer specified a technically unnecessary job requirement (e.g. a university degree for a manual job), which would statistically work against applicants from a particular racial or ethnic group. Other countries are often less explicit, including race in anti-discrimination legislation that covers gender, sexuality and disability. Anti-discrimination law is seen as a democratic minimum, but also as a way to integrate racial and ethnic minorities by providing a minimum level of protection. It is nevertheless often up to the individual to bring a case against discrimination.

Multiculturalist legislation and approaches are more contentious than anti-racism, because they can be perceived to go against the grain of values of 'fairness', understood as equal treatment for all, which is a key value in liberal democracies (Smith 2012).

KEY DEBATES ABOUT MULTICULTURALISM

- To what extent should a liberal democracy institute differential treatment for recognised categories (such as ethnic groups) in the public sphere; does this advance or retard equality? That is, do equality and fairness mean a right to difference or a right to indifference (i.e. being treated without regard to racial or cultural difference)? Would differentialist policies lead to segregation and more inequality?
- Can difference be confined to the 'private' sphere – something people simply enact in the home – or does the private always intrude into the sphere deemed to be 'public' – as religious practices often do – meaning it should be recognised there too (Modood and Werbner 1997)? For example, if 'private' religious and ethnic differences are the object of racist discrimination, doesn't that automatically make them a 'public' matter?

There are no simple answers to these questions (Fraser and Honneth 2003; Modood 2007; Taylor and Gutmann 1994). They reflect basic tensions at the heart of liberal democracies (see Chapter 3): how do you create equality in a society where differences (of race, ethnicity, culture, gender, etc.) are the basis for inequality?

Britain has tended towards some recognition of multiculturalist difference, as part of an attempt to actively manage 'race relations' – a category not admitted to exist in France. These relations were perceived to be in crisis in the wake of 1981 'race' riots in several British cities, as British black and Asian youth reacted against marginalisation, harassment from neo-fascist groups and racism, especially in the police. In the education system, for example, policy recommendations have, since the 1980s, emphasised the need for minority students to feel that their ethnic identities are protected and valued, partly as a response to evidence that black and some, but not all, Asian students were doing less well in school. Although by 2000 the government had begun to fund some Muslim schools, alongside other state-funded faith schools (Modood and May 2001), in practice, multiculturalist actions – often the subject of intense public debate – have ranged from serving halal meat and scheduling single-sex PE lessons to learning about 'diversity' in Britain. More recently there have been calls to emphasise the 'core values' that all people living in Britain share, in order to avoid 'sleepwalking' into segregation, as Trevor Phillips, the black British head of the Commission for Racial Equality, put it in 2005. His comments came in the wake of the July 2005 London terrorist bombings, which were a motive for some people to re-think the place given to supporting minority cultural identity, and especially Islam. This was fomented by the fact that cities such as Leicester were on the brink of having majority non-white populations. Meanwhile, survey research was showing that a large majority of British Muslims identified first as Muslims and only second as British; they also tended to be averse to secular and integrationist attitudes (Schain 2012: 20). Multiculturalism didn't seem to have done the trick.

In France, the state rejected British-style multiculturalism and is officially blind to race and ethnicity. In contrast to Britain, Muslims there were much more likely to identify with France first and Islam second, and to value integration (Schain 2012: 20). Perhaps powerfully assimilationist policies worked better. In practice, the state recognises and negotiates with immigrant associations and, increasingly, Islamic organisations, now that Muslim youths are seen as a 'problem'. The state also channels special funds to Educational Priority Zones (ZEPs), which are defined in part by the proportion of immigrants in them. But France has banned the public wearing of full-face veils by Muslim women and the use of headscarves (and indeed any 'conspicuous' religious symbol) in schools (see below), whereas in Britain the full-face veil is not (yet) banned and schools decide their own dress codes (Bleich 2003; Schain 2012).

All this adds up to a conditional kind of integration promoted by citizenship, anti-racist and multiculturalist policies: the overarching sense – more so in France than in Britain – is that immigrants are a problem that needs to be addressed.

Racial inequality, racial violence, nation, class, diaspora

Whatever the mechanisms of integration, racial inequality remains a reality, although one that exists alongside very marked class inequality within the white population.

RACIAL AND ETHNIC INEQUALITY IN BRITAIN AND FRANCE

Racial and ethnic inequalities have multiple causes in addition to racial discrimination: for example, immigrants arriving with low qualifications and skills will enter the job market at a low level. The data below show both simple inequalities and direct racial discrimination.

For Britain:

- In 2010 poverty rates for seven-year-old children were nearly 75% for Pakistani and Bangladeshi children, over 50% for black children, and 25% for white children.
- In 2010 unemployment rates were 7% for whites, 9% for Indians, and 13–17% for Pakistani, Bangladeshi, Chinese, black and mixed-race people. In 2012, for males aged sixteen to twenty-four, unemployment was 23% for blacks and 13% for whites (Institute of Race Relations 2014).
- Research conducted in British cities shows that job applicants with identifiably 'ethnic' names suffered 29% more rejection than applicants with 'British' names who had similar qualifications (Wood et al. 2009). This said less about discrimination against Afro-Caribbean and black British people, whose names could not be reliably linked to ethnic or racial origins.
- In 2008–9 black people were about 3% of the population, but were 15% of those stopped by the police (Equality and Human Rights Commission 2011: 134).

In France official statistics are not organised by race or ethnicity, but by national origin.

- For non-EU immigrants unemployment in 2010 was 23% compared to 9% for non-immigrants; for Africans it was 35–37% (Observatoire des inégalités 2011).
- In 2011 the average family income for people born of African immigrants was 30% lower than for people of French parents (Lombardo and Pujol 2011).
- A 2006 study showed a 'traditional French' job applicant was nearly four times more favoured than an Arab or black African applicant when short-listing, and six times more favoured at the interview stage (Observatoire des inégalités n.d.).
- A 2009 study of Parisian train stations found that black people were seven to twelve times and Arab people two to fifteen times more likely than whites to be stopped by the police (Open Society Justice Initiative 2009).

Alongside the raw facts of racial inequality and evidence of racial discrimination in the job market and the criminal justice system, several other factors militate against integration. Although over 40 per cent of people in Britain and France think that immigrants are integrating well, the majority think they are not, indicating that immigrants are seen as outsiders (Schain 2012: 19). An extreme symptom of this is the racist violence and expressions that are common all over Europe, often overlapping with religious hatred and targeting Jews and Muslims, as well as non-whites. As Werbner (2002: 18) says, 'racial violence is the opposite of everyday ethnicity or multiculturalism'. Specific racist murders, such as that of black teenager Stephen Lawrence in London in 1993, have assumed iconic status. In England and Wales the police registered over 37,000 racially or religiously aggravated crimes in 2011–12. In France – where data are recorded very differently, making direct comparisons difficult – there were over 1,000 racist incidents reported in 2009, plus 815 anti-Semitic incidents (Camus 2011; Institute of Race Relations 2014).

In both France and Britain right-wing anti-immigrationist political parties have enjoyed some electoral success and, even if many people reject some of the far-right organisations as neo-fascist and racist – a 'lunatic fringe' – these parties tap into a wider view that immigration has 'gone too far', that some immigrants get 'unfair' advantages from a multi-culturalist state that is ignoring native white communities, especially poorer ones. These more pervasive views are linked to underlying ideas of the nation as European, white and Christian. As Gilroy argued for Britain, the sense that the country has been in a post-colonial decline since WWII is articulated through a pervasive antipathy to immigrants seen as outsiders and as creating 'problems' and 'threats'. This antipathy is only partly offset by the emerging cosmopolitan conviviality that he detects in certain multiracial areas of London (Gilroy 1987, 2004) and that others have identified in parts of Leicester (McLoughlin et al. 2014).

Conviviality may work in some areas of London, but in others the sense of decline and marginalisation is dominant. In Bermondsey, east London, white working-class people feel that they have been bypassed both by a class structure that relegates them to the margins, as it does immigrant minorities, and by a state multiculturalism that they feel offers no space to white people – unless they can produce a viable 'culture' – and gives them no reason for making common cause with the immigrants who live alongside them. Such alienation does not necessarily ally these people with the far right, but it does suggest they feel dislocated in a locality that was once properly *theirs* and in which they perceive non-white people as still outsiders (Evans 2010).

SOLIDARITY SOUP?

In Paris in 2004 a group called Solidarité des Français, linked to the right-wing political movement Bloc Identitaire, ran a soup kitchen serving only 'identity soup', with pork in it, expressly to exclude Muslims (not to mention Jews). The idea spread to Nice and to Belgium. The French authorities eventually, in 2007, banned the action as discriminatory. Odile Bonnivard, the president of Solidarité des Français, declared that in France outsiders were being given precedence over native French people; she justified her action with the right-wing nationalist slogan 'Our own before the others'.

Minority reaction

In the face of this complex and varied scenario, immigrant minorities and their descendants have reacted in equally varied ways, also depending a good deal on the origin of the immigrants (Ali, Kalra and Sayyid 2006; Ali 2003; Centre for Contemporary Cultural Studies 1982; Fryer 1984; Hussain and Bagguley 2012; McLoughlin et al. 2014). In Britain Afro-Caribbean immigrants and their children have tended to integrate quite extensively, despite racism and discrimination, but in the process have also transformed (urban) British society. Caribbean island-based identities generally coalesced in the 1970s into a broader 'black' identity, which also began to overlap, in uneven ways, with the identifications of young British Asians. Black expressive culture, in the form of music, dance and events such as the Notting Hill Carnival in London, both conveyed a sense of contestatory identity, challenging marginalisation and racism, and became part of the mainstream British youth culture. Asian immigrants and British Asians are often seen as having integrated less, but this varies a great deal by class and origin: Bangladeshi and Pakistani immigrants of poorer backgrounds tend to be more segregated. Again, cities such as Bradford, with a 27 per cent Asian population, or Leicester with 35 per cent, have been transformed by this presence: it is difficult to think simply in terms of 'integration' into a 'host' society here; instead, new urban social and cultural forms have been generated (McLoughlin 2006; McLoughlin et al. 2014).

An expression of these new forms is residential segregation, a product of complex interactions between processes of exclusion and minority strategies to cope and forge communities and identities, as we saw for the United States. In Europe religious values (Islamic, Hindu, Sikh) are often an important source for forging such community identities, which are de facto racialised in the local context. For Britain and France racial segregation certainly exists – more so in Britain than France – and, although it has shown some decline over time, it has been slow and not very marked for most groups.

ARE THERE GHETTOS IN FRANCE AND BRITAIN?

Wacquant (2008) concluded that, in comparison to the United States, the Parisian *banlieue* (peripheral housing estates) were characterised by the 'anti-ghetto'. The non-white communities there were not racially homogeneous, having a mixture of Arab, black African, black Caribbean and Asian residents, plus often a majority of white French people; the communities had porous boundaries, with more successful people moving out, little shared cultural identity, and not much in the way of local institutions that would support a collective identity. The strongly assimilationist culture of France had, at least so far, blocked the use of ethnoracial categories as a major frame for social perceptions and relations.

Lapeyronnie (2012) came to the opposite conclusion: the *banlieue* riots of 2005 – not mentioned by Wacquant – were evidence that poverty, unemployment, poor access to education and especially racism had created levels of exclusion and isolation that resembled the classic ghetto, in which residents had formed an alternative society, with its own cultural norms, an informal economy and a collective sense of being non-white. Ghetto boundaries shaped social networks, creating strong ties within the ghetto and weak ones across its boundaries, and also marked social behaviour: women donned the headscarf within the ghetto, conforming to gendered ideas about female modesty, and often removed it when going outside.

people who value community spirit, quietness, propriety and respectability. There are class differences within the village – not everyone can afford the expensive golf club membership – but local people do not (yet) feel seriously menaced by the urban 'underclass' lifestyles that they think characterise poorer areas of Leicester city. Rural Leicestershire and urban Leicester are opposed in their minds, and in the local government tourist publicity for the province: the former is typically English, symbolised by iconic foods such as local pork pies and Stilton cheese; the latter is exotically urban and cosmopolitan, with areas dominated by Asian restaurants and gold jewellery shops. Race is not mentioned, but is linked to place and food consumption – as we saw in Chapter 2, food often carries racialised meanings (Slocum and Saldanha 2013).

Asians are present in the village in two ways: as local shopkeepers and as wealthy residents. The shopkeepers experienced some explicit harassment when they moved in – racist graffiti painted on their walls – but the people Tyler spoke with described the Asians in positive, non-racist terms as polite, hard-working and providing a very good service. At the same time, Asians were consistently marked as other. Of course they were seen as not white and not English, but this was not often mentioned explicitly. Instead, difference was marked in two main ways. First, the shopkeepers were said to get ahead by marshalling the labour of all family members, regardless of age and gender, into the enterprise. Asian men were thought to subordinate the women and children in the family in ways implicitly marked as traditional and perhaps backward, in contrast to the supposedly more egalitarian and modern gender and generational relations of white village residents.

Second, the shopkeepers were perceived to succeed not just by hard work – a valued quality – but by 'staying open all hours' and by charging 'a bit extra'. This implied an excessive quality, going beyond what was considered normal. Asian shopkeepers were seen to have displaced English merchants, and Asians would displace more English people from the job market in future. The English were portrayed by some villagers in self-deprecating fashion, as honest and hard-working, but a bit naïve – 'we are still wringing bloody mops out and wearing knotted handkerchiefs on our heads going down Skegness' (a traditional but declining seaside holiday resort in England) (2012: 58). The ambitious and hard-working Asians were overtaking the English, but partly by dint of excess – exploiting their own family members, working too many hours, charging too much, etc.

The theme of excess and otherness also emerged when Greenville villagers talked about the wealthy Asians who lived in expensive houses in the village. They were described as isolating themselves with fences and electric gates, not socialising in the village, working all the time, making ostentatious additions to their houses – including in one case a prayer room like a 'mosque' – to accommodate multiple family members. There were fears, which proved groundless, that Asian families would run businesses from their houses. One man disparagingly described a white neighbour of an Asian family as a racist, but then sympathised with the fact that 'he has got Indians living to the back and side of him', with what he imagined that implied in terms of 'three families moving into one house' and 'multiples of kids running round the garden' (2012: 70).

Tyler emphasises the fact that middle-class white Britons hold racist views, even as they attribute racism to others and praise Asian villagers as hard-working and polite. She counters the common academic view that working-class white people are the main locus of racism in Britain, often theorised as these people's reaction to the state multiculturalism that supposedly 'unfairly' benefits minorities. Underlining this, she presents data from Coalville, an ex-mining town near Leicester, 99 per cent white and mostly working class, where Asians are present as shopkeepers and restaurant-owners. The town was undergoing rapid economic change away from the mining industry towards services and light industry, and Tyler argues that these changes also gave younger people there a sense of changing attitudes and values. They criticised the 'traditional' racist views of some of their parents and grandparents, describing them as linked to a former era. Racism was linked to other 'traditional' views such as sexism: fathers and grandfathers were described as more likely to be both sexist and racist. Younger people acknowledged that some of their own generation could be racist too, but they often challenged such views, in some cases drawing on personal family histories of immigration (e.g. from Poland), which de-racialised Asian immigration by arguing that white and non-white immigrants were similar. Such a colour-blind argument acknowledges that all immigrants face challenges and exclusions, but it also ignores the impact of racism and erases the colonial dimensions of Asian immigration.

Tyler also argues that, despite their critical intentions, these younger people also implicitly built on ideas of non-whites as others: they took it for granted that Asians were shopkeepers and owned take-away food outlets; it was assumed that a 'big Black fella' might be scary to other people. Even these everyday anti-racists, then, dealt in familiar racial stereotypes and silenced colonial histories. In addition, basic values of morality, fairness and propriety informed white working-class views. So-called problem families – white families seen as welfare scroungers, idlers, drug-users, etc. – were considered 'unfair' beneficiaries of the state, displacing deserving families from government housing. Asians, as people who supposedly also benefited 'unfairly', became associated with deviance from values of propriety and fairness. Discourses about class were thus racialised to sustain a sense of racial difference between whites and Asians.

Tyler's final location was 'Streetville', an inner-city neighbourhood in Leicester, which was 47% Asian or British Asian, 30% white and 13% black, and with a mainly semi- and unskilled working population. In an account reminiscent of the studies of New York and Chicago cited in Chapter 7, Tyler looks at a local residents' association (RA), which participated in a broader forum, made up of black, BrAsian and multi-ethnic community organisations, which lobbied to defend the local area. The RA was 60 per cent white and was a space in which white residents could express resentment about Asians. Some white members complained that Asians did not participate (they self-segregated), that they took time off neighbourhood business for Ramadan (they were excessively religious) and that some Asian youths vandalised the neighbourhood (they were deviant). They also complained that Asians excluded whites from their community centres and that local Bangladeshis were 'unfairly' getting a new state-funded community centre. Such views were challenged by others in the RA, both white and Asian, who explained what Ramadan meant, challenged the allegations of vandalism, clarified that community centres were open to all, offered space in an Asian centre to the RA, and proposed lobbying the city council for more space for the RA (rather than simply blaming the Asians). These measures could not by criteria of parental origin and place of residence. The meanings of blackness were also shaped, however, by what W. E. B. Du Bois (1897) called 'double consciousness', that is, the constant awareness of oneself through the eyes of (exclusionary) others and the strategies of individual and collective identification used in coping with that. Black youths were always aware that they were being judged by white people, who were likely to view them through powerful stereotypes, and they tended to construct and perform their identities with that in mind – hence the book's title, *The art of being black*. One result of that double consciousness was that, while black identity was in some ways multiple and contested – especially in relationships between black people – in other ways it was presented by the young men as straightforward, a simple fact of life defined by not being white.

Alexander's second book focused again on men, and this time on the emergence of the image of the Asian gang in London in the 1990s. Asian culture had previously been, and still could be, seen by British mainstream society as strong, traditional, law abiding and a positive resource for building community – which could however also be resistant to 'integration'. Ideas about Asian culture were focused largely on women, with concerns about patriarchal oppression, arranged marriages and enforced norms of modesty. With second- and third-generation youngsters going through school, the idea of young Asian men being a 'problem', linked to poverty, crime, violence and racial conflict, began to circulate, and was expressed in the press and government reports as an issue about ethnically based 'gangs'. This built on existing ideas about young black men being 'in crisis' (Cashmore and Troyna 1982) and carrying out street muggings (Hall et al. 1978).

At the same time, and pre-9/11, Islam was already emerging as a source of concern for the British authorities and liberal commentators. The protests surrounding the publication of Salman Rushdie's *Satanic Verses* (1988), in which Muslims worldwide accused Rushdie of blasphemy, and which led to a public book-burning in Bradford in 1989, indicated the more visible presence of Islamic militancy. In this sense, Muslims have 'become the new "black", with all the associations of cultural alienation, deprivation and danger that come with this position' (Alexander 2000: 15). Young Asian men were caught in an intersection of ideas about ethnic minorities, males and youths as potential problem areas.

Against this backdrop, Alexander did ethnographic research with the South Asian Youth Organisation (SAYO) on the 'Stoneleigh Estate' in south-east London, an area that was about 31 per cent black and Asian. She worked with teenagers and young men who described themselves as Bengali, although they were born in Britain. SAYO was established by local authorities as a way to include Asian youths, seen as having been sidelined by existing provision and being the subject of racist exclusion by local whites: the plan was to channel resources and support to them via a youth club. Perhaps predictably, social services were accused by local mothers of 'favouritism' towards 'certain races', while Asian youngsters were accused of intimidating other young people (2000: 64). In fact, the Bengali young men told of histories of racial harassment and violence in the area, which had lessened once the young boys had grown up a bit and learned to stand up for themselves.

Alexander shows how the anti-Asian racism on the estate and in its schools did not attract very much attention, despite involving attacks by firebombing and groups of white men armed with knives and baseball bats – although the move to create SAYO was one response. In contrast, when a series of fights and conflicts between young Asians and young black men occurred in the local school (which was 80 per cent black), this was quickly seized upon by the schools, the police and local government as being an issue of racial gang violence that needed immediate attention. The fights were actually driven by quite specific and personal rivalries, and the need the boys felt to save face, not back down, assert their masculinity, avenge specific wrongs and be loyal to friends in the peer group. The fights were not 'racial' as such – in the sense that Asians did not automatically target all blacks or vice versa – and the boys denied that race was behind the conflicts. They also criticised one of their number who attacked a white boy apparently just because he was white, although they later defended him in the court case that followed the attack.

But the SAYO boys spoke about how the black boys – or some of them – were trying to dominate the school space, and how Asians had to fight for their own space and reputation. They also saw racial–ethnic divisions as obvious and indeed natural: one said, 'It's just that all the black youths stay together, all the Bengalis like stay together'; another argued that some of the Bengali boys found it 'hard to mix with other races' and a third boy said that it was 'natural' for his group to be 'just Bengali' (2000: 100, 142). In that sense race was a real 'absent-presence' (2000: 3). It was present in the way the conflicts were identi-fied by police and local authorities as specifically racial as well as gang driven (2000: 107); it was denied by the boys as a specific motive for action; yet it was present in the way they thought about a – loosely bounded and non-exclusive – collective control of space and sense of belonging, which operated at a very local level, was articulated through specific relationships and rivalries, and was intimately connected to masculinity.

The naturalness of the Bengali peer group was a complex articulation of belonging to a locality, an age-set and a gender, and of sharing a set of experiences. The peer group appeared tight and solidary, but was actually full of shifting alliances and, over time, was quite ephemeral, not lasting much beyond school. The groups also had complex relations with other groups, which oscillated between friend and foe: a lot of the fights the boys had were with other Asians, again around personal conflicts.

Alexander argues that Asian youths have in the past been seen by social scientists, and especially anthropologists, as caught 'between two cultures' – a problematic approach, as noted above. More recently they have been seen as exponents of the 'new Asian cool', a fashionable set of hybrid musical and other consumable cultural products. She wants to avoid both approaches and reassert the importance of racial inequality and power differences: tracing the way 'Asian youth have become synonymous with crisis and with threat' – as exemplified by the figure of 'the Asian gang' – is a good means of doing this (2000: 226).

8.6 Geographies of race in black Liverpool

The studies by Tyler and Alexander alert us to the importance of place in ideas about race. The village, the urban neighbourhood, the school and, more widely, the nation are all examples of the varied places that shape race. Race never stands alone, but is always in

Brown's study is useful for a number of reasons: it brings in a longer-term history, which helps to relativise the image of 'race in Britain' as being a post-WWII phenomenon; it draws out the importance of place for race, but understands both as being constituted in diasporic mode; its focus on race mixture draws attention to gendered dynamics in the constitution of local racialised identities; and the delineation of disagreements about blackness illustrates the unevenness of the reproduction of racial difference.

8.7 Algerians in France

Race is not a term or concept that is often deployed in France. It is not unusual in Europe today for the term to appear mainly in the context of anti-racist discourse (see Chapter 5), but this tendency is particularly marked in France. Even black Antillean French people (citizens from the Caribbean French departments of Martinique and Guadeloupe) do not use the term much in talking about their situation – although they may talk about racism as a problem – and French workers, and even more so professionals, do not refer to race or skin colour much when discussing issues of community boundaries and distinctions between 'us' and 'them' (Beriss 2004a, 2004b; Lamont 2004).

This is partly because of the dominant ways in which culture, race and nation are conceived in France (Beriss 2004b; Silverstein 2004). On the one hand, there is a powerful republican, universalist, assimilationist model of the nation-state, which includes within the boundaries of French culture those who choose to join the political community of French citizens and participate in its public language and institutions. This concept has been built, since the late eighteenth century, against the idea of race as bio-cultural difference, for which skin colour has operated as a prime marker. Room may be left here for a 'right to difference' – as French President Mitterrand declared in 1981 – but such difference has to be superficial and domesticated (i.e. tamed and private).

On the other hand, there is a powerful concept of French culture as a specific life-way, rooted in French history, located in French territory, based on Catholicism and drawing on rural peasant traditions. This strongly naturalised and more exclusive concept of French culture defines those perceived as culturally different as outside the nation, unless they can assimilate culturally. Thus black immigrants and citizens are not rejected by French workers because they are black, although they may be because they are culturally different: of course, skin colour can act as a signal of cultural difference and, as black Antilleans are aware, many white French people cannot distinguish by sight between a French citizen from Martinique and a Congolese immigrant. But it is immigrants from North Africa and their French-born children who are seen as more culturally different. In interviews French workers described them as lacking a work ethic and getting 'unfair' amounts of state resources, especially in education; they were seen as lacking in civility – rude, lacking respect – oppressive towards women, tending towards criminality and violence, and overly influenced by Islam; above all, they were seen as unable or unwilling to assimilate (Lamont 2000, 2004).

Paul Silverstein's ethnography of Algerians in France powerfully conveys this sense of difference and the way Algerian immigrants and their French-born offspring – often known as *beurs* – cope with it (Silverstein 2004). First, he makes clear that, alongside the universalist model of French citizenship, the French state has also recognised and institutionalised cultural difference, whether by promoting Berber ethnicity in Algeria during the colonial period, or by declaring the 'right to difference' in the 1980s. His book shows how difference operates in Paris in ways that are simultaneously cultural, religious and racial: Algerians are seen as culturally alien, visibly non-white and Muslim. Extreme expressions of this have included racist violence, for example in the 1973 'summer of death' in which 150 Algerian workers were attacked, and 15 killed, around Marseilles by groups of French men who used a language of 'rat-hunting' to describe their activities. In the *cités* (public housing projects) of Paris and Lyons in the 1980s immigrants' children were shot at by French residents using .22-calibre hunting rifles, while police and security guards used excessive force against North Africans in supermarkets and train stations in the *cités*, leading to several deaths.

Silverstein's first focus is on the *cités* located in the *banlieues* (working-class suburbs on the periphery of Paris), where many immigrants and their children live. Although white French working-class people are usually in the majority in this public housing, the *cités* are routinely presented in the media as dominated by non-whites, as well as being trouble spots and sites of urban decline, unemployment, social deprivation, crime and violence. Racial difference is given a powerful spatial expression, with Paris itself being mainly white, despite its colourful spots where immigrant culture is visible and can be consumed by tourists and the middle classes, while the *banlieues* and especially the *cités* are the location of the non-white immigrant working-class masses, located there by state housing policies. These areas are connected to Paris by trains – which run from centre to periphery, but barely connect different peripheral locations – with distance from Paris being popularly perceived to correlate to increased danger from crime and violence.

Within the *banlieues* there were internal subdivisions. Pantin, a municipality of 60,000 people, where Silverstein lived, was divided by a canal into Le Petit Pantin and Quatre-Chemins, with the latter having more immigrants, more public housing projects, a much poorer and more North African feel in its markets and shops, and fewer cultural resources. When a public gym was temporarily moved from Le Petit Pantin to Quatre-Chemins local Algerian and black youths started to use it, with the approval of the gym manager. Soon conflicts started to develop over the place being left in a mess, petty theft, vandalism, etc. The manager shouted at the boys and, later, talking to Silverstein, said: 'I'm not a racist, but those youth, those Blacks and North Africans, always mess things up for us.' One of the boys, of Algerian parents, said, 'He's completely racist. We have as much right to be there as he does. We live here' (2004: 102). The conflict was about making a mess in the gym, but also about where the gym really belonged spatially and thus racially: it belonged to Le Petit Pantin, a whiter space where certain standards of behaviour were the norm, but it was temporarily located in Quatre-Chemins, a more non-white immigrant space, where slightly different behaviours held sway.

Silverstein emphasises that, in the *cités*, sites of spatial isolation and economic exclusion, immigrants build alternative lifestyles based on values that are 'not necessarily isomorphic with those projected by assimilation and integration projects of the French nation-state'. They do so in part by importing Algerian objects and decorations into their apartments and, more generally, inhabiting 'multiple spaces between France and Algeria simultaneously' (2004: 78–9). These translocal, diasporic connections create new racialised spaces within the French nation.

Silverstein explores sport and religion as domains where racial and ethnic difference occurs. On the one hand, football is an arena in which a national, inclusive Frenchness is articulated, which can glorify difference in the shape of star players such as Zinedine Zidane, backed by corporate sponsors such as Nike. This acknowledges difference, but assimilates it to the French nation. The French state also made sport a key plank in its regeneration projects in the *cités* aimed at countering the threat of radical Islam: this was perceived to be running rampant through the *banlieues* in the context of the 1990s civil war in Algeria, which pitted Islamist groups against the Algerian government and which led to bombings in Paris and Lyons. Alongside increased surveillance and police round-ups, the state provided sports facilities – routinely trashed by locals during confrontations with the police – and sports programmes, including summer camps, which were designed to compete with Islamic summer camps and were promoted by sports stars of North African parentage and corporate sponsors such as Nike.

ADVERTISING THE DIFFERENCE

A 1995 advertising campaign by Nike played on the theme of religious and racial difference with these slogans:

* 'There is not but one God; there are eleven [i.e. the national football team]'. This referenced the Muslim belief that 'There is but one God and his name is Allah'.
* 'No law prohibits you from wearing the PSG [football team Paris Saint-Germain] jersey to school'. This referenced legislation regulating the wearing of 'conspicuous' religious symbols – aimed at the Muslim headscarf – to school (Silverstein 2004: 126).

The message was that unity through sport trumps divisive religion, but it implied that Islam was not properly French.

In the domain of religion, the French state first built a mosque in 1921 to honour Algerian fighters in WWI; it did not function as a house of worship, however. In the 1950s employers such as Renault and Citroen were allowed to create prayer rooms for Muslim workers. This was a tactic of divide and rule – to undermine unions – which slightly backfired when unionised Muslim workers demanded more prayer rooms in the 1970s. In the wake of Mitterrand's 1981 declaration of a 'plural France', where people had a 'right to difference', associations of immigrant and *beurs*, which included many Muslim associations, proliferated and were recognised by the state, partly as places where the second generation could keep off the streets. By the 1990s, when fears of transnational Islamic terrorism were stoked by the involvement of *beurs* in bombings on French soil, such associations were also subjected to surveillance and police con-trol – and countered by state sports programmes, which were in turn countered by sports programmes run by Islamic associations. The state tried, and failed, to define an 'Islam for France', working with Muslim organisations to define an acceptable form of Islam, free from the political agendas that supposedly characterised most Islam and that offended against the French principle of *laïcité* (secularism), which insisted on divorcing religion from politics. After the 9/11 attacks of 2001 the state renewed its efforts, creating the French Council of the Muslim Faith, an independent and elected body, which, in theory, had to respect *laïcité*.

THE HEADSCARF: GENDER, RELIGION AND RACE

The debates about the Muslim headscarf, already mentioned above, reveal the gendered workings of these overlapping racial and religious identifications (Silverstein 2004: 139–47). For men the defining sartorial symbols of racial–religious difference are white knitted skullcaps and long white robes. These may single out the wearer for police attention, but they have not spurred any public debate, as they are not usually worn by young men at school. Young men, in contrast, tend to adopt the '*beur* look', wearing the clothes that Nike and other sportswear companies promote, which resonates with US inner-city African American styles, and which one newspaper warned might turn out to be the uniform of the next generation of the 'soldiers of Islam' (2004: 149). Alarmism aside, these mainstream consumerist styles have not provoked debate. In contrast, the headscarves and veils worn by women have caused huge public debates. This is because women figure as a key locus for identification and the marking and bounding of difference (see Chapters 2 and 3). They are seen as the bearers of 'culture' and tradition, as well as the literal bearers and nurturers of the next generation; access to and control by men of women, their sexuality and their reproductive powers are typically seen as key aspects of the maintenance – or the destruction – of group boundaries and identities (Nagel 2003; Yuval-Davis 1997).

In France, and elsewhere, the headscarf or veil, because it is worn by women, has been represented in multiple ways, by both Muslim and non-Muslim women (and men), and related to both gender and ethnic–racial hierarchies. Some see it as a positive statement of a religious identity, which is also a post-colonial affirmation of ethnic–racial identity; others see it as a perverse rejection of the opportunity to integrate into French society on equal terms; some see it as an emblem of women's oppression by patriarchal norms of modesty and sexual control; others see it as an expression of women's autonomy, freeing them from an objectifying sexual gaze and allowing them some access to public spaces. Others still point out that the entire debate ends up reinforcing the role of women as key boundary figures in identity politics, and their role as 'problem' against an apparently neutral background (Bracke and Fadil 2011).

In France, as in some other countries, the trend has been towards tighter regulation. Guidelines in 1989 allowed schools to make their own decisions about students wearing 'ostentatious' religious symbols, but by 1994 the steer from government was firmer and in practice 150 girls were expelled in 1994–5 for wearing the scarf. In 2003 the headscarf was banned in schools and, in 2010, any full-face veil worn in public (see Fig. 8.2).

Silverstein's ethnography, and other works on France cited here, are useful in bringing to the fore the issue of the racialisation of religion. In Britain this has become an issue too, but the fact that North African immigration has from an early date been a major factor on the French scene highlights the way religious difference has articulated with race.

Figure 8.2. Muslim woman in Paris; the text on her placard reads, 'The veil over your eyes is much more dangerous than the veil on my hair'.
(Directphoto Collection/Alamy)

CONCLUSION

France and Britain are often seen as having taken different paths across the tricky terrain of immigration, multiculturalism, citizenship and nation. France has had its republican assimilationist model, in which official multiculturalism has been downplayed and, above all, race has had very muted presence. In Britain the management of 'race relations' has been the official line, with a more US style of multiculturalist pluralism in which different groups are allowed separate institutional spaces. In fact, the two countries also have a great deal in common, along with many European countries. Popular culture – football, music, food, etc. – shows the huge impact of fifty years or so of intensive post-colonial immigration. Mixed marriages are relatively common in both countries. Diasporic connections have linked both places to multiple spaces around the world: transnational spaces have emerged for politics and culture, which bridge Europe, Africa and Asia (not to mention Latin America). In important ways both societies have been transformed in this time, dismantling the old idea of a host society and its immigrant minorities, which either assimilate or do not. Over fifty years many white British and French people have changed their ideas about what their societies are like, and are prepared to accept multicultural and multiracial nations.

Yet racial inequality and substantial segregation remain and, while they may be declining slowly, are still major factors. One may quibble about whether ghettos really exist in the Parisian *banlieues*, but even if statistically the segregation is much less than in the United States, the experiences of non-whites living there invite comparisons with US cities. Immigration is still a very difficult and contentious area: more and more complex technology is being deployed to control it, yet humanitarian and economic crises in Africa and the Middle East are forcing rapidly increasing numbers of asylum seekers and desperate economic migrants to cross the Mediterranean – the distinction between political and economic motives begins to fade here. European countries have little option but to accept many of these migrants – albeit arguing about how various countries should share the cost – while deporting others.

Support for right-wing anti-immigrationist parties is growing, but this hides the fact that every political party has to deal with the thorny issue of immigration when the electorate routinely estimates that the number of migrants in the country is two to three times the actual number, where a large minority or a majority reckons there are too many immigrants in the country, and where large majorities are concerned about illegal immigration (German Marshall Fund of the United States 2013: 40). Of course, concerns about immigration are not always about non-white immigration. In Britain Romanians and Bulgarians have been encountering increased hostility since they became EU citizens in 2007 and were able, in theory, to move freely within the EU – in fact, several countries imposed a seven-year period of restrictions on their entry and job-seeking. Their numbers in Britain in 2013 were a mere fifth of immigrants from India, yet there were scares about 'hordes' and 'invasions'. The overall concern is about who will do what in the nation – access welfare services, housing and jobs, or commit crimes, etc. – and race becomes woven, in varying ways, into those anxieties, sometimes disappearing but often being implicitly or explicitly present. Overall, then, there is no doubt that immigration and the idea of the nation as fundamentally white continue to be the key frame for understanding race in Europe.

FURTHER ACTIVITIES

1. There are a number of useful websites, with lots of data on citizenship, immigration and multiculturalism. These allow you to make detailed comparisons between different countries.
   a. EUDO Observatory on Citizenship (http://eudo-citizenship.eu/).
   b. Migrant Integration Policy Index (http://www.mipex.eu/).
   c. Multiculturalism Policy Index (http://www.queensu.ca/mcp/immigrant.html).

2. Check out the website of the European Agency for the Management of Operational Cooperation at the External Borders of the Member States of the European Union (Frontex) (http://frontex.europa.eu/). Explore how they talk about the risks the EU faces from illegal migration.

3. The novel *Kilo* by Mohammed Yunas Alam (2002) gives a good account of the English city of Bradford – sometimes known as Bradistan – from the point of view of a young British Asian man.

4. Research the controversies over headscarves in France and Britain, to see similarities and differences in how the issue has been handled in each country.

9 Conclusion

We have come on a long and varied journey in this book, covering a long time span from the ancient Greeks and Chinese through to the twenty-first century, and drawing examples from many geographical locations, including India, Japan, Australia, the Americas, Africa and Europe. My aim has been to analyse what race is – or what we should take it to be analytically – and how it works by demonstrating the varied forms that thinking about race and racial difference can take, and the different ways racial exclusion and inclusion can operate. In conclusion, I summarise some key ideas by addressing two areas in the theorising of race: (a) what race is; and (b) how we can explain its presence and operation. I then explore some dimensions of race as a global phenomenon.

9.1 Theorising race

How have social scientists tried to explain race? In essence, once they had concluded that race was not a biological reality, and was a 'social construction', explaining it became a matter of relating racial categories and racial thinking to other social processes and ideas. As we saw in Chapter 1, central to these other factors have been coloniality (encompassing colonialism, African slavery and post-colonial power differences) and capitalism: race has widely been seen as driven by the politics and economics associated with these processes, past and present.

Although race has been theorised in relation to other social forces, many scholars have argued that race should not be 'reduced' to being a side-effect of these forces (see Chapters 4 and 6): as a set of ideas and practices embedded in social contexts, race can drive other processes as well as being driven by them (Omi and Winant 1986). As we have seen, race can indeed become a major force, with the power to, among other things, explain 'everything', as Robert Knox put it in 1850 (see Chapter 3); to create what Du Bois called 'double consciousness' (see Chapter 8); to produce radical forms of segregation, complete with an elaborate etiquette of behaviour and rights in the US South and South Africa (see Chapter 7); and to shape ideas about the nation, development and immigration (see Chapters 6 and 8).

Race has its own 'materiality', which is not just the physical signs of race, but the way ideas and practices around race become embedded in the material fabric of society (M'charek 2013). Although anthropologists have been important in the theorisation of ethnicity (Banks 1996), they have been less influential than sociologists and philosophers in the theorisation of race (Back and Solomos 2000; Essed and Goldberg 2002; Harrison 1998; Mukhopadhyay and Moses 1997), having been first embarrassed by the role early anthropology played in sustaining scientific racism and then remaining somewhat trapped in the, admittedly important, question of the non-existence of biological race. But, as I have shown in Chapters 6–8, anthropologists have been important in tracing the materiality of race in everyday life.

What is race? Embeddeness and relationality

If the preceding chapters have shown one thing, it is that race is incredibly varied in its manifestations. In Chapter 1 I defined race in terms of the intersection of two sets of criteria: a set of ideas that categorise difference in terms of bodies, blood, heredity, filiation, nature and behaviour (ideas that can be found in many contexts that for analytic purposes we do not want to include under the umbrella of race); and a set of practices and ideas linked to Western colonial domination and its consequences worldwide. This gives us an idea of the range of phenomena we are dealing with, always bearing in mind that concepts such as body, blood, heredity and nature need to be understood in their local and historical context. But the key point that emerges from the historical and more recent examples in this book is that race is embedded in specific social contexts and draws its meanings and effects from them, as it also shapes those contexts. This is a corollary of the idea that race is a 'social construction': ideas and practices around race exist in relation to and embedded in other social practices and ideas. Race never stands alone, but is an integral part of ideas and practices, usually related to other forms of difference, and usually those of power and inequality. Being integral means race is not simply shaped by other, more primary forces. Instead, we need to think of race as a series of meanings and practices that are a constitutive part of a social formation, *alongside* other elements, rather than being a superstructure to more 'basic' drivers.

There are various conceptual tools that help us to think about this embeddedness and relationality. One is the idea of a network or assemblage of connected elements (ideas, practices), in which certain sets of relationships have, with repeated enactments over time, become tight, dense and relatively stable, without being fixed (Latour 2005; Ong and Collier 2008). New connections and conjunctures can always be made – for example, a fishing net can be twisted into new shapes – but will not necessarily become stable unless they are enacted repeatedly by people. Another conceptual tool is that of articulation, as developed by Stuart Hall (Grossberg 1996). This also posits a constellation of ideas and meanings, which have been worked into certain relationships, forming discourses that tend to be attached to particular people, although not in a fixed way. For example, elements of Christianity may be articulated together, along with other, apparently disparate elements, and articulated to Rastafarians as subjects, producing a religious discourse of political resistance. Some discourses become hegemonic, that is, powerful, widely agreed upon and taken for granted by many. The word 'articulation' nicely conveys both the idea of relationality (two bones are articulated in a joint) and of expression or discourse.

A further conceptual tool for thinking about the relationality of race is that of topology (M'charek, Schramm and Skinner 2014b). Topology is a branch of mathematics that studies the properties of space that are preserved under continuous deformations, including stretching and bending. A subway map is a topological map as it shows the stable relationships of a network of subway stations, whether the map is deformed into a sphere, a square or a long rectangle. The image of the subway map is limited, however, because points in the network are always in stable relationships. Used in a more metaphorical way, with less emphasis on fixed points, topology suggests the way ideas and meanings can be twisted into new conjunctures, which may retain some of the relationships between the ideas or may create new relationships.

The material on race in Europe can be thought about using these tools. Tyler's work shows how components such as 'rural village', 'Englishness', 'fairness', 'pork pies' and 'whiteness' – to mention only a few – have been brought into a tight but flexible relationship with each other. They connote each other, but they do not all always have to be present, so that 'whiteness' does not have to be explicit, for example, but can be connoted by 'Englishness' or 'rurality': unlike a subway map, you do not always have to go through the subway station of 'white' as you travel between 'rural' and 'English'; the map can be twisted to conjoin the latter two, with whiteness left as a trace or an absent presence. The relationships between all these elements work in a particular way for Greenville, but they can work at a variety of scales, including the national. They have the status of common sense; they are taken for granted and are relatively stable, although not incontestable. 'Blackness' or 'non-whiteness' can also be related to a host of other meanings and practices, such as 'immigrant', 'foreign language', 'strong food smells', 'noise', 'Islam', 'cool', 'youth', 'multiculturalism', 'changing times', 'progress' and 'decline' – again, to name only some. As before, the way these elements are articulated – that is, brought into relationship with each other and expressed in words or actions – is highly flexible: they can connote each other, without all elements necessarily being present. Ideas about the decline of Britain can thus invoke a series of ideas about immigration, non-whiteness and multiculturalism, without these being explicit. For example, in the wake of victories for right-wing parties in the European elections of May 2014, the leader of France's National Front, Marine Le Pen, said 'our people' wanted 'French politics by the French, for the French, with the French' (Willsher 2014: 5). She was talking about resisting control by European Union legislators, not about limiting non-white immigration, but the question of race was there between the lines, because of the way 'our people' has figured in related articulations by her and others, which exclude 'alien' immigrants from the nation.

Some articulations or constellations of meanings have great traction because they articulate familiar ideas. In Britain there was a public furore in June 2014 about Muslim groups allegedly involved in a 'Trojan horse' plot to impose an Islamist agenda on some schools in Birmingham. The original allegation may have been a hoax, but it triggered a series of government enquiries, mostly driven by Michael Gove, the secretary of state for education, who in 2006 had published a book, *Celsius 7/7*, warning of the threat to Britain posed by Islamic extremism and terrorism. A government report claimed there was a conspiracy to impose 'a hardline and politicised strand of Sunni Islam' on some schools (Clarke 2014).

Government inspectors suddenly decided that one school, Park View, was in need of special measures, as it was doing too little to 'raise students' awareness of the risks of extremism' (Ofsted 2014). Park View School officials – both Muslim and non-Muslim – strongly denied the report and claimed that they were the object of a witch hunt. An independent report concluded there had been improper attempts to influence educational provision in some schools but 'no evidence of a conspiracy to promote radicalisation or violent extremism' (Trojan Horse Review Group 2014). The details of the incident are complex, but the point is how easily narratives, which articulated elements about Muslims, threats to British values, Trojan horses, indoctrination of youth, religious conservatism and extremist terrorism, could gain traction in the public sphere. As usual, race was not mentioned explicitly, but was tacitly present, because of the way the elements 'Muslim', 'outsider', 'immigrant', 'Britain/Europe' and 'non-whiteness' have frequently been articulated in existing and historical discourses.

These examples all point to the way race figures in relational assemblages involving the nation. Just as important is the relation of race to class. We saw in Chapters 5 and 6 in particular how structural inequalities of class form a network of meanings and practices in which race is embedded and which race also shapes. This means that people can pursue economically rational and apparently neutral behaviours – maintaining property values, sending their children to good schools, promoting regional development, avoiding risky neighbourhoods, minimising awkward interactions in university residences – without being intentionally racist and without apparent reference to race. But, because blackness is relationally and multiply linked to poverty and low status, these behaviours do discriminate against black people and in the end are racist – even if structurally or institutionally so – because they rely on the perception of blackness as usually linked to low status.

We have also seen in several chapters how race figures in a relational network with gender and sex. This is obvious in Latin America, where the narrative of *mestizaje* always invokes, at some level, the image of sex between light-skinned dominant men and dark-skinned subordinate women, quintessentially the domestic servant. Talk of sex, in a specific context, can thus easily be a way of referencing race without mentioning it as such. The same goes for the United States, where talk of rape or sexual impropriety could carry a racialised subtext, because the issue of interracial sex was hedged about with so much taboo. This is not to say that any mention of gender and sex immediately connotes race in these regions – just as any mention of poverty does not always immediately connote blackness. But the relational matrix that articulates these racial, sexual and class meanings is such that they can connote each other by means of a topological deformation in which certain elements can drop out of a particular articulation, without disappearing altogether. To talk of *la muchacha* (the girl, or the servant) connotes the possibility of racial difference in many Latin American contexts when the overt reference is to gender and age subordination; on the other hand, to talk of *la chola* (the urbanised indigenous woman, but also often the maid) is more explicitly racialised and potentially connotes both class subordination and sexual availability.

The embeddedness of race makes it a protean and slippery phenomenon. At the end of Chapter 1 I quoted Mosse's idea that racism is a 'scavenger ideology' (Mosse 1985: 234).

This captures some of the truth, but it portrays racism as a ready-made ideology that goes around finding elements that will nurture and support it. In fact, racism is also a parasitic shape-shifting ideology, which adapts itself to other elements – class, religion, gender, sports – as it also inflects their mode of operation. The answer to the question 'what is race?' can be answered in part by the combination of the two criteria I mentioned above – a type of categorisation and a specific history of coloniality – but we always have to be attentive to the relational way race is embedded in social contexts and takes on particular forms. Hence the difference between the forms race took in the seventeenth and the nineteenth centuries and between Latin America, the United States and Europe.

Why racism and race?

Why do race and racism exist as ideology and practice? They are both linked to practices of exploitation and inequality – in that sense, it is like asking 'why does inequality exist?' – but the question is really about why exploitation, inequality and exclusion take the particular form of using racial categorisations, in all their variety. The simple answer is that an appeal to the complex of ideas that link bodies, blood, heredity, filiation, nature and behaviour provides a very powerful and flexible toolkit for thinking about difference and for 'naturalising power' (Yanagisako and Delaney 1995). This toolkit, or parts of it, can be used to organise exclusion, exploitation and violence in situations that I would not class as racial (e.g. ancient Greece, ancient China, late twentieth-century Bosnian–Serb conflicts). Race, in my view, refers to uses of the toolkit in the context of European coloniality.

   The point is that a toolkit that links behaviour to ideas about bodies, filiation and human nature (bearing in mind that concepts of human nature vary over time and space) has a powerful elective affinity with practices of exclusion and hierarchisation. Throughout the variety that I have traced – in relation to the categories of people constituted by European coloniality – we can see the linking of behaviour to bodies (internal and/or external) that are connected through filiation, in ways that render behaviour natural. This provides an effective and flexible way of thinking about difference and a basis for exclusion and discrimination. It is by no means a rigid or clear basis, because bodies and natures are frequently understood as unstable formations subject to change, as we have seen – hence the anxieties about degenerating constitutions of Europeans in early twentieth-century Dutch colonies (see Chapter 3). But it is an effective and powerful basis.

9.2 Globalising race

Race is by definition a globalising ideology in several respects. First, racial thinking from early on was a theory about humanity as a whole and its internal diversity: racism, while highly exclusionary, has a universalist dimension (Balibar 1991b). Nineteenth-century racial science aspired to universal generalisations about humanity; and arguments from genetics, which mostly propose that no such thing as races exist biologically, but some of which contend that biological divisions akin to familiar 'races' do exist, are also universalist theories, put forward by global science (see Chapter 4). Universalism also is reflected in the 'cultural fundamentalism' – sometimes called 'new racism' or 'neo-racism' – that contends that it is universal human nature to want to be with 'one's own kind', whether 'kind' is interpreted only in terms of culture or also in terms of descent and appearance (see Chapter 5).

Second, the fact that race relies on the conceptual toolkit of naturalising categorisations outlined above means that racial ideologies find easy resonance in many areas of the world. Ideas of difference similar in many ways to racial thinking existed in China and Japan before extensive encounters with the West and, when Western racial ideas did make an impact on these countries in the nineteenth century, they were easily grafted on to existing ideas (see Chapter 3; Dikötter 1992, 1997b; Takezawa 2005).

Third, and most important, the roots of race in Western colonialism and capitalism mean that racial concepts had a global reach, driven by the increasing global dominance of whiteness and the gradually more encompassing categories of non-whiteness that signified a lower rung on the globalising political–economic ladder. This ladder always had its complexities – in the form of poor whites and wealthy non-whites – and since decolonisation it has steadily become more complex and multiple, and will continue to do so, especially as economies in the global South (China, Brazil, India, etc.) compete effectively with Europe and the United States. Yet in a very broad sense global inequalities still have racial dimensions. This is evident in patterns of migration from poorer to richer regions, whether this is from Africa to Europe, from Mexico to the United States or from South Asia to the Middle East oil states. As we saw in Chapter 8, the European Union has lists specifying which countries' citizens need and do not need a visa to enter the EU, and these 'have the effect of distinguishing between white and nonwhite entrants to the EU' (M'charek, Schramm and Skinner 2014b: 474).

The racialisation of the global political economy is also evident in the structure of exploitative and informal industries, which target the vulnerable. In sex tourism the majority of the traffic is white men from the global North going to have sex with non-white women of the global South (South East Asia, the Caribbean, Brazil, etc.) (Sánchez Taylor 2010). Human trafficking, which is very largely about the sex trade too, is more complex, as there is a lot of intra-regional trafficking of women and children, but people trafficked into Europe and North America are largely from the global South (despite the growing participation of some Eastern European countries) (US Department of State 2013). The global traffic in human organs also shows racialised patterning: 'In general, the flow of organs follows the modern routes of capital: from South to North, from Third to First World, from poor to rich, from black and brown to white, and from female to male' (Scheper-Hughes 2000: 193).

The transnational dimensions of race are not only about Western ideas and practices being imposed on other regions. Race has diasporic dimensions that work in a more circulatory and dialogic way. Afro-Brazilian intellectuals, who travelled to West Africa and England in the nineteenth century and returned to their country armed with ideas about Yoruba culture, influenced the way blackness was conceived in Brazil. Liverpool-born blacks find diasporic points of reference in histories of interaction with the United States and Africa (see Chapters 6 and 8). African Americans visit Nigeria on heritage tours that feed into the culturalisation of concepts of blackness in the United States, while others take DNA ancestry tests, which trace their genetic connections to West Africa, thus adding biological data to ideas of race (Clarke and Thomas 2006; Nelson 2008; Schramm 2012).

Pierre traces how concepts of race in Ghana have been shaped by long-standing interactions between Ghana, Britain and the United States: while race is often seen as the territory of the United States and the African diaspora and not relevant to most of independent Africa, Pierre argues that 'race is the modality' through which many identifications of class, nationality, ethnicity and gender within Ghana are structured; indeed, 'The very production of "Africa" … occurs through ideas of race' (Pierre 2013: 5). Loci of transnational interaction include British colonial categories of race and the blackface concert parties of the early 1900s, intended for white audiences but seen by Africans too; the transnational pan-Africanism of Kwame Nkrumah, Ghana's socialist leader; the more recent presence of white development and peace corps workers; and African American cultural influences, ranging from heritage tourism to films and music, which feed changing images of blackness in the United States, but also shape Ghanaian ideas of blackness. US blacks are hailed as brothers and sisters, whose hoped-for economic solidarity does not live up to expectations: a transatlantic sense of blackness is thus at once created by claims of kinship and fractured by economic disappointment.

Pierre's work shows the historical depth of these transnational circulations, but it is clear that recent globalisation has intensified diasporic possibilities. Thomas describes 'modern blackness' in Jamaica as being 'urban, migratory, based in youth-oriented popular culture and influenced by African American popular style, individualistic, radically consumerist and ghetto feminist' (2004: 241). For young working-class Jamaicans, modern blackness partly displaces but still exists alongside the 'respectability' of the middle classes and nation-building elites, derived from colonial hierarchies and valuing lightness of skin, school education, moderation, female domesticity, hard work and the Anglican church. Modern blackness builds on but departs from the traditional working-class idea of 'reputation', which values egalitarianism, physical prowess, sexual conquest, verbal fluency, manual skills and practical knowledge, and the rum shop (Wilson 1973). In the Caribbean – long an epicentre of global circulation – this respectability/reputation opposition was already transnationally constituted. Migration to the United States and the influence of African American consumer culture creates new forms that have multiple valencies: modern blackness valorises the dollar, but also resents US dominance of Jamaica and the racism that Jamaican migrants endure in the United States.

9.3 The future of race

The examples above indicate that racial difference and racialised inequality are persistent aspects of today's world. While racial segregation may be slowly diminishing in the United States, and unevenly so in the United Kingdom and perhaps France, there is little evidence that racialised difference as a whole is diminishing. Given the relationality and embeddedness of race, it is optimistic to expect that social inequality – which has increased in many countries, North and South, over the last thirty years – will not be

Banks, Marcus. 1996. *Ethnicity: anthropological constructions*. London: Routledge.

Banton, Michael. 1987. *Racial theories*. Cambridge: Cambridge University Press.

Barbary, Olivier. 2004. El componente socio-racial de la segregación residencial en Cali. In *Gente negra en Colombia: dinámicas sociopolíticas en Cali y el Pacífico*, edited by Olivier Barbary and Fernando Urrea, 157–94. Cali and Paris: CIDSE/Univalle, IRD, Colciencias.

Barbary, Olivier, Héctor F. Ramírez, Fernando Urrea et al. 2004. Perfiles contemporáneos del la población afrocolombiana. In *Gente negra en Colombia: dinámicas sociopolíticas en Cali y el Pacífico*, edited by Olivier Barbary and Fernando Urrea, 69–112. Cali and Paris: CIDSE/Univalle, IRD, Colciencias.

Bardaglio, Peter W. 1999. 'Shamefull matches': regulation of interracial sex and marriage in the South before 1900. In *Sex, love, and race: crossing boundaries in North American history*, edited by Martha Hodes, 112–38. New York: New York University Press.

Barkan, Elazar. 1992. *The retreat of scientific racism: changing concepts of race in Britain and the United States between the world wars*. Cambridge: Cambridge University Press.

Barker, Martin. 1981. *The new racism: Conservatives and the ideology of the tribe*. London: Junction Books.

Barnett, Steve. 1976. Coconuts and gold: relational identity in a south Indian caste. *Contributions to Indian Sociology* 10(1): 133–56.

Bayly, Susan. 1995. Caste and race in the colonial ethnography of India. In *The concept of race in south Asia*, edited by Peter Robb, 165–218. Delhi: Oxford University Press.

——— 1999. *Caste, society and politics in India from the eighteenth century to the modern age*. Cambridge: Cambridge University Press.

Bedoya, Gabriel, Patricia Montoya, Jenny García et al. 2006. Admixture dynamics in Hispanics: a shift in the nuclear genetic ancestry of a South American population isolate. *Proceedings of the National Academy of Sciences of the United States of America* 103(19): 7234–9.

Beriss, David. 2004a. *Black skins, French voices: Caribbean ethnicity and activism in urban France*. Boulder: Westview Press.

——— 2004b. Culture-as-race or culture-as-culture: Caribbean ethnicity and the ambiguity of cultural identity in French society. In *Race in France: interdisciplinary perspectives on the politics of difference*, edited by Herrick Chapman and Laura L. Frader, 111–40. Oxford: Berghahn.

Bernstein, Nell. 2003. Goin' gangsta, choosin' cholita. In *Signs of life in the USA: readings on popular culture for writers*, edited by Sonia Maasik and Jack Solomon, 599–604. Boston: Bedford Books and St Martin's Press.

Birchal, Telma S., and Sérgio D. J. Pena. 2011. The biological nonexistence versus the social existence of human races: can science instruct the social ethos? In *Racial identities, genetic ancestry, and health in South America: Argentina, Brazil, Colombia, and Uruguay*, edited by Sahra Gibbon, Ricardo Ventura Santos and Mónica Sans, 69–99. New York: Palgrave Macmillan.

Blackburn, Robin. 1998. *The making of New World slavery: from the baroque to the modern, 1492–1800*. London: Verso.

Bleich, Erik. 2003. *Race politics in Britain and France: ideas and policymaking since the 1960s*. Cambridge: Cambridge University Press.

——— 2011. *The freedom to be racist? How the United States and Europe struggle to preserve freedom and combat racism*. New York: Oxford University Press.

Bliss, Catherine. 2009. Genome sampling and the biopolitics of race. In *A Foucault for the 21st century: governmentality, biopolitics and discipline in the new millennium*, edited by Samuel Binkley and Jorge Capetillo, 320–37. Boston: Cambridge Scholars Publishing.

Boas, Franz. 1912. Changes in the bodily form of descendants of immigrants. *American Anthropologist* 14(3): 530–62.

——— 1966 [1940]. *Race, language and culture*. New York: Free Press.

Bonilla-Silva, Eduardo. 2003. *Racism without racists: color-blind racism and the persistence of racial inequality in the United States*. Lanham, MD: Rowman & Littlefield.

Bonnett, Alistair. 2000. *Anti-racism*. London: Routledge.

Borges, Dain. 1993. 'Puffy, ugly, slothful and inert': degeneration in Brazilian social thought, 1880–1940. *Journal of Latin American Studies* 25(2): 235–56.

Bracke, Sarah, and Nadia Fadil. 2011. 'Is the headscarf oppressive or emancipatory?' Field notes on the gendrification of the 'multicultural debate'. *Religion and Gender* 2(1): 36–56.

Brah, Avtar. 1996. *Cartographies of diaspora: contesting identities*. London: Routledge.

Brennan, Jonathan, ed. 2002. *Mixed race literature*. Stanford: Stanford University Press.

Brockington, John. 1995. Concepts of race in the Mahabharata and Ramayana. In *The concept of race in South Asia*, edited by Peter Robb, 97–108. Oxford: Oxford University Press.

Brodkin, Karen. 1998. *How Jews became white folks and what that says about race in America*. New Brunswick: Rutgers University Press.

Brown, Jacqueline Nassy. 2005. *Dropping anchor, setting sail: geographies of race in Black Liverpool*. Princeton: Princeton University Press.

Brown, Ryan A., and George J. Armelagos. 2001. Apportionment of racial diversity: a review. *Evolutionary Anthropology* 10: 34–40.

Buffon, Georges. 1807. Of the varieties in the human species. In *Buffon's Natural History containing a theory of the earth, a general history of man, of the brute creation, and of vegetables, minerals, etc.*, edited by J. S. Barr, 190–352. London: T. Gillet.

Burchard, Esteban Gonzalez, Elad Ziv, Natasha Coyle et al. 2003. The importance of race and ethnic background in biomedical research and clinical practice. *New England Journal of Medicine* 348(12): 1170–5.

Burdick, John. 1998. *Blessed Anastácia: women, race, and popular Christianity in Brazil*. London: Routledge.

2013. *The color of sound: race, religion, and music in Brazil*. New York: New York University Press.

Caldwell, Kia Lilly. 2007. *Negras in Brazil: re-envisioning black women, citizenship, and the politics of identity*. New Brunswick: Rutgers University Press.

Campbell, Ben. 2007. Racialization, genes and the reinventions of nation in Europe. In *Race, ethnicity and nation: perspectives from kinship and genetics*, edited by Peter Wade, 95–124. Oxford: Berghahn.

Camus, Jean-Yves. 2011. *Racist violence in France*. Brussels: European Network against Racism.

Candelario, Ginetta E. B. 2007. *Black behind the ears: Dominican racial identity from museums to beauty shops*. Durham: Duke University Press.

Canessa, Andrew. 2012. *Intimate indigeneities: race, sex and history in the small spaces of Andean life*. Durham: Duke University Press.

CARA. 2011. *Guidelines governing the adoption of children, 2011*. Central Adoption Resource Authority [cited 26 February 2014]. Available from http://adoptionindia.nic.in/guideline-family/Overwiew.html.

Cárdenas, Roosbelinda. 2012. Multicultural politics for Afro-Colombians: an articulation 'without guarantees'. In *Black social movements in Latin America: from monocultural mestizaje to multiculturalism*, edited by Jean Muteba Rahier, 113–34. New York: Palgrave Macmillan.

Carens, Joseph H. 2003. Who should get in? The ethics of immigration admissions. *Ethics & International Affairs* 17(1): 95–110.

Carter, Bob, Clive Harris and Shirley Joshi. 2000. The 1951–55 Conservative government and the racialisation of black immigration. In *Black British culture and society: a text reader*, edited by Kwesi Owusu, 23–39. London: Routledge.

Cartmill, Matt. 1998. The status of the race concept in physical anthropology. *American Anthropologist* 100(3): 651–60.

Edwards, Jeanette. 2000. *Born and bred: idioms of kinship and new reproductive technologies in England*. Oxford: Oxford University Press.

Edwards, Jeanette, and Carles Salazar, eds. 2009. *European kinship in the age of biotechnology*. Oxford: Berghahn.

Eliav-Feldon, Miriam, Benjamin Isaac and Joseph Ziegler, eds. 2009. *The origins of racism in the West*. Cambridge: Cambridge University Press.

Ellison, George T. H., Andrew Smart, Richard Tutton et al. 2007. Racial categories in medicine: a failure of evidence-based practice? *PLoS Medicine* 4(9): e287.

Eltis, David. 1987. *Economic growth and the ending of the transatlantic slave trade*. Oxford: Oxford University Press.

Eltis, David, and David Richardson. 2010. *Atlas of the transatlantic slave trade*. New Haven: Yale University Press.

Engerman, Stanley L., and Kenneth L. Sokoloff. 2005. The evolution of suffrage institutions in the New World. *Journal of Economic History* 65(4): 891–921.

Entine, Jon. 2001. The straw man of 'race'. *World and I* 16(9): 294.

　　2008. *Taboo: why black athletes dominate sports and why we're afraid to talk about it*. New York: PublicAffairs.

Epstein, Steven. 2007. *Inclusion: the politics of difference in medical research*. Chicago: University of Chicago Press.

Equality and Human Rights Commission. 2011. *How fair is Britain? Equality, human rights and good relations in 2010*. London: Equality and Human Rights Commission.

Erasmus, Zimitri. 2005. Race and identity in the nation. In *State of the nation: South Africa 2004–2005*, edited by John Daniel, Roger Southall and Jessica Lutchman, 9–33. Cape Town and East Lansing, MI: HSRC Press and Michigan State University Press.

Escobar, Arturo. 2008. *Territories of difference: place, movements, life, redes*. Durham: Duke University Press.

Essed, Philomena, and David Theo Goldberg, eds. 2002. *Race critical theories: text and context*. Oxford: Blackwell.

Evans, Gillian. 2010. 'What about white people's history?': class, race and culture wars in twenty-first-century Britain. In *Culture wars: contexts, models and anthropologists' accounts*, edited by Deborah James, Plaice Evelyn and Christina Toren, 115–35. Oxford: Berghahn.

Eze, Emmanuel Chukwudi, ed. 1997. *Race and the enlightenment: a reader*. Oxford: Blackwell.

Fair Housing Center of Greater Boston. 2006. *The gap persists: a report on racial and ethnic discrimination in the Greater Boston home mortgage lending market*. Boston: Fair Housing Center of Greater Boston.

　　n.d. *Historical shift from explicit to implicit policies affecting housing segregation in Eastern Massachusetts*. Fair Housing Center of Greater Boston [cited 14 April 2014]. Available from http://www.bostonfairhousing.org/timeline/index.html.

Fanon, Frantz. 1968. *The wretched of the earth*. New York: Grove Press.

　　1986 [1952]. *Black skin, white masks*. London: Pluto Press.

Fausto-Sterling, Anne. 2000. *Sexing the body: gender politics and the construction of sexuality*. New York: Basic Books.

Fernandez, Nadine T. 2010. *Revolutionizing romance: interracial couples in contemporary Cuba*. New Brunswick: Rutgers University Press.

Fields, Barbara J. 1982. Ideology and race in American history. In *Region, race, and reconstruction: essays in honor of C. Vann Woodward*, edited by J. Morgan Kousser and James M. McPherson, 143–77. Oxford: Oxford University Press.

Fischer, Kirsten. 2002. *Suspect relations: sex, race, and resistance in colonial North Carolina*. Ithaca: Cornell University Press.

FitzGerald, David Scott, and David Cook-Martín. 2014. *Culling the masses: the democratic origins of racist immigration policy in the Americas*. Cambridge, MA: Harvard University Press.

Fluehr-Lobban, Carolyn. 2005. *Race and racism: an introduction*. Walnut Creek, CA: AltaMira Press.

Fogg-Davis, Hawley. 2002. *The ethics of transracial adoption*. Ithaca: Cornell University Press.

Foucault, Michel. 1998. *The will to knowledge. The history of sexuality: Volume 1*. Translated by Robert Hurley. London: Penguin.

   2003. *'Society must be defended': lectures at the Collège de France, 1975–1976*. Translated by David Macey. New York: Picador.

Fox, Jonathan, and Gaspar Rivera-Salgado, eds. 2004. *Indigenous Mexican migrants in the United States*. San Diego, CA: Center for US–Mexican Studies, Center for Comparative Immigration Studies, University of California at San Diego.

Frankenberg, Ruth. 1993. *White women, race matters: the social construction of whiteness*. London: Routledge.

Fraser, Nancy, and Axel Honneth. 2003. *Redistribution or recognition? A political–philosophical exchange*. London: Verso.

Frederickson, George M. 2002. *Racism: a short history*. Princeton: Princeton University Press.

French, Jan Hoffman. 2009. *Legalizing identities: becoming black or Indian in Brazil's northeast*. Chapel Hill: University of North Carolina Press.

Fry, Peter. 2000. Politics, nationality, and the meanings of 'race' in Brazil. *Daedalus* 129(2): 83–118.

Fryer, Peter. 1984. *Staying power: the history of black people in Britain*. London: Pluto Press.

Fryer, Roland G., Jr. 2007. Guess who's been coming to dinner? Trends in interracial marriage over the 20th century. *Journal of Economic Perspectives* 21(2): 71–90.

Fujimura, Joan H., Troy Duster and Ramya Rajagopalan. 2008. Introduction: race, genetics, and disease: questions of evidence, matters of consequence. *Social Studies of Science* 38(5): 643–56.

Garrido, Margarita. 2005. 'Free men of all colours' in New Granada: identity and obedience before Independence. In *Political cultures in the Andes, 1750–1950*, edited by Cristóbal Aljovín de Losada and Nils Jacobsen, 165–83. Durham: Duke University Press.

German Marshall Fund of the United States. 2013. *Transatlantic trends: key findings 2013*. Washington, DC: German Marshall Fund of the United States.

Gheera, Manjit, and Robert Long. 2013. *Inter-racial adoption*. London: House of Commons Library, Social Policy Section.

Gies, Frances, and Joseph Gies. 2010. *Marriage and the family in the Middle Ages*: New York: HarperCollins.

Gilman, Sander L. 1985. *Difference and pathology: stereotypes of sexuality, race, and madness*. Ithaca: Cornell University Press.

Gilroy, Paul. 1987. *'There ain't no black in the Union Jack': the cultural politics of race and nation*. London: Hutchinson.

   1993. *The black Atlantic: modernity and double consciousness*. London: Verso.

   2000. *Between camps: nations, cultures and the allure of race*. London: Penguin.

   2004. *After empire: melancholia or convivial culture*. London: Routledge.

Glacken, Clarence J. 1967. *Traces on the Rhodian shore: nature and culture in Western thought from ancient times to the end of the eighteenth century*. Berkeley: University of California Press.

Goldberg, David Theo. 1993. *Racist culture: philosophy and the politics of meaning*. Oxford: Blackwell.

   2008. *The threat of race: reflections on racial neoliberalism*. Malden, MA: Wiley-Blackwell.

Holt, Thomas C. 1992. *The problem of freedom: race, labor, and politics in Jamaica and Britain, 1832–1938*. Baltimore: Johns Hopkins University Press.

Hooker, Juliet. 2009. *Race and the politics of solidarity*. Oxford: Oxford University Press.

Horsman, Reginald. 1981. *Race and manifest destiny: the origins of American racial Anglo-Saxonism*. Cambridge, MA: Harvard University Press.

Howell, Signe. 2003. Kinning: the creation of life trajectories in transnational adoptive families. *Journal of the Royal Anthropological Institute* 9(3): 465–84.

2006. *The kinning of foreigners: transnational adoption in a global perspective*. Oxford: Berghahn.

Howell, Signe, and Marit Melhuus. 2007. Race, biology and culture in contemporary Norway: identity and belonging in adoption, donor gametes and immigration. In *Race, ethnicity and nation: perspectives from kinship and genetics*, edited by Peter Wade, 53–71. Oxford: Berghahn.

Htun, Mala. 2004. From 'racial democracy' to affirmative action: changing state policy on race in Brazil. *Latin American Research Review* 39(1): 60–89.

Huet, Marie-Hélène. 1993. *Monstrous imagination*. Cambridge, MA: Harvard University Press.

Hume, David. 1987. *Essays, moral, political, and literary*, edited by Eugene F. Miller. Indianopolis: Library of Economics and Liberty.

Humes, Karen R., Nicholas A. Jones and Roberto R. Ramirez. 2010. *Overview of race and Hispanic origin: 2010*. Washington, DC: US Bureau of the Census.

Hunter, Margaret L. 2002. 'If you're light you're alright': light skin color as social capital for women of color. *Gender and Society* 16(2): 175–93.

Hussain, Yasmin, and Paul Bagguley. 2012. *Riotous citizens: ethnic conflict in multicultural Britain*. Farnham: Ashgate.

Ifekwunigwe, Jayne O. 1999. *Scattered belongings: cultural paradoxes of 'race', nation and gender*. London: Routledge.

Ifekwunigwe, Jayne O., ed. 2004. *'Mixed race' studies: a reader*. London: Routledge.

Institute of Race Relations. 2014. *Statistics*. Institute of Race Relations [cited 29 May 2014]. Available from http://www.irr.org.uk/research/statistics/.

International HapMap Consortium. 2003. The International HapMap Project. *Nature* 426(6968): 789–96.

Isaac, Benjamin H. 2004. *The invention of racism in classical antiquity*. Princeton: Princeton University Press.

Jacob, Margaret C. 1988. *The cultural meaning of the scientific revolution*. Philadelphia: Temple Univerisity Press.

Jefatura del Estado. 1988. Ley 35/1988, de 22 de noviembre, sobre Técnicas de Reproducción Asistida. *Boletín Oficial del Estado* 282: 33373–8.

Johnson, Kevin R. 1998. Race, the immigration laws, and domestic race relations: a 'magic mirror' into the heart of darkness. *Indiana Law Journal* 73(4): 1111–59.

Johnson, Lyman L., and Sonya Lipsett-Rivera, eds. 1998. *The faces of honor: sex, shame, and violence in colonial Latin America*. Albuquerque: University of New Mexico Press.

Jordan, Winthrop. 1977. *White over black: American attitudes toward the Negro, 1550–1812*. New York: Norton.

Kahn, Jonathan. 2013. *Race in a bottle: the story of BiDil and racialized medicine in a post-genomic age*. New York: Columbia University Press.

Kalra, Virinder S., ed. 2009. *Pakistani diasporas: culture, conflict, and change*. Karachi: Oxford University Press.

Kaszycka, Katarzyna A., and Jan Strzałko. 2003. 'Race': still an issue for physical anthropology? Results of Polish studies seen in the light of the US findings. *American Anthropologist* 105(1): 116–24.

Katzew, Ilona. 2004. *Casta painting: images of race in eighteenth-century Mexico*. New Haven: Yale University Press.

Katzew, Ilona, and Susan Deans-Smith, eds. 2009. *Race and classification: the case of Mexican America*. Stanford: Stanford University Press.

Kaur, Raminder, and Virinder S. Kalra. 1996. New paths for South Asian identity and musical creativity. In *Dis-orienting rhythms: the politics of the new Asian dance music*, edited by Sohan Sharma, John Hutnyk and Ashwani Sharma, 217–31. London: Zed.

Kearney, Michael. 2000. Transnational Oaxacan indigenous identity: the case of Mixtecs and Zapotecs. *Identities – Global Studies in Culture and Power* 7(2): 173–95.

Kelley, Robin D. G. 1997. *Yo' mama's disfunktional! Fighting the culture wars in urban America*. Boston: Beacon Press.

Kent, Michael, and Peter Wade. 2015. *Social Studies of Science*, in press. Genetics against race: science, politics and affirmative action in Brazil. Unpublished work. Manchester: University of Manchester.

Kevles, Daniel J. 1995. *In the name of eugenics: genetics and the uses of human heredity*. 2nd edn. Cambridge, MA: Harvard University Press.

Knox, Robert. 1850. *The races of men: a fragment*. London: Henry Renshaw.

Koenig, Barbara A., Sandra Soo-Jin Lee, and Sarah S. Richardson, eds. 2008. *Revisiting race in a genomic age*. New Brunswick: Rutgers University Press.

Kolchin, Peter. 2002. Whiteness studies: the new history of race in America. *Journal of American History* 89(1): 154–73.

Krimsky, Sheldon, and Kathleen Sloan. 2011. *Race and the genetic revolution: science, myth, and culture*. New York: Columbia University Press.

Kuper, Leo. 1965. *An African bourgeoisie: race, class and politics in South Africa*. New Haven: Yale University Press.

Lamont, Michèle. 2000. *The dignity of working men: morality and the boundaries of race, class, and immigration*. Cambridge, MA: Harvard University Press.

2004. Immigration and the salience of racial boundaries among French workers. In *Race in France: interdisciplinary perspectives on the politics of difference*, edited by Herrick Chapman and Laura L. Frader, 141–61. Oxford: Berghahn.

Lamont, Michèle, Mario Luis Small and David J. Harding. 2010. Introduction: reconsidering culture and poverty. *Annals of the American Academy of Political and Social Science* 629(1): 6–27.

Lapeyronnie, Didier. 2012. *Ghetto urbain: ségrégation, violence, pauvreté en France aujourd'hui*. Paris: Robert Laffont.

Larson, Brooke, Olivia Harris and Enrique Tandeter, eds. 1995. *Ethnicity, markets and migration in the Andes: at the crossroads of history and anthropology*. Durham: Duke University Press.

Latour, Bruno. 1993. *We have never been modern*. Translated by Catherine Porter. London: Harvester Wheatsheaf.

2005. *Reassembling the social: an introduction to actor-network-theory*. Oxford: Oxford University Press.

Leibbrandt, Murray, Christopher Woolard and Ingrid Woolard. 2009. Poverty and inequality dynamics in South Africa: post-apartheid developments in the light of the long-run legacy. In *South African economic policy under democracy*, edited by Janine Aron, Brian Kahn and Geeta Kingdon, 270–99. Oxford: Oxford University Press.

Leibbrandt, Murray, Ingrid Woolard, Arden Finn et al. 2010. *Trends in South African income distribution and poverty since the fall of apartheid*. OECD Social, Employment and Migration Working Papers 101. Paris: OECD Publishing.

Lentin, Alana. 2004. *Racism and anti-racism in Europe*. London: Pluto.

2014. Post-race, post politics: the paradoxical rise of culture after multiculturalism. *Ethnic and Racial Studies* 37(8): 1268–85.

Lentin, Alana, and Gavan Titley. 2011. *The crises of multiculturalism: racism in a neoliberal age*. London: Zed.

Lévi-Strauss, Claude. 1963. *Totemism*. Boston: Beacon Press.

Levine, Lawrence. 1977. *Black culture and black consciousness: Afro-American folk thought from slavery to freedom*. New York: Oxford University Press.

Lewis, Bernard. 1992. *Race and slavery in the Middle East: an historical enquiry*. Oxford: Oxford University Press.

Lewis, Laura A. 2012. *Chocolate and corn flour: history, race, and place in the making of 'black' Mexico*. Durham: Duke University Press.

Lewis, Oscar. 1959. *Five families: Mexican case studies in the culture of poverty*. New York: New American Library.

Lewontin, Richard C. 1972. The apportionment of human diversity. *Evolutionary Biology* 6: 381–98.

Liberman, Anatoly. 2009. *The Oxford etymologist looks at race, class and sex* [cited 27 August 2011]. Available from http://blog.oup.com/2009/04/race-2/.

Licops, Dominique. 2002. Redefining culture, politicizing nature: negotiating the essentialism of natural metaphors of identification in the work of James Clifford, Paul Gilroy and Aimé Césaire. *Thamyris/Intersecting: Place, Sex and Race* 8(1): 53–68.

Lieberman, Larry, and R. C. Kirk. 2002. The 1999 status of the race concept in physical anthropology: two studies converge. *American Journal of Physical Anthropology Supplement* 34: 102.

Lieberman, Leonard, and Larry T. Reynolds. 1996. Race: the deconstruction of a scientific concept. In *Race and other misadventures: essays in honor of Ashley Montagu in his ninetieth year*, edited by Larry T. Reynolds and Leonard Lieberman, 142–73. Dix Hills, NY: General Hall Inc.

Linebaugh, Peter, and Marcus Rediker. 2000. *The many-headed hydra: sailors, slaves, commoners, and the hidden history of the revolutionary Atlantic*. London: Verso.

Linke, Uli. 1999. *Blood and nation: the European aesthetics of race*. Philadelphia: University of Pennsylvania Press.

Livingstone, Frank B., and Theodosius Dobzhansky. 1962. On the non-existence of human races. *Current Anthropology* 3(3): 279–81.

Lockhart, James, and Stuart Schwartz. 1983. *Early Latin America: a history of colonial Spanish America and Brazil*. Cambridge: Cambridge University Press.

Lombardo, Philippe, and Jérôme Pujol. 2011. *Le niveau de vie des descendants d'immigrés*. Paris: Insee Références.

Lovell, Peggy. 1994. Race, gender and development in Brazil. *Latin American Research Review* 29(3): 7–35.

Loveman, Mara. 2009. Whiteness in Latin America: measurement and meaning in national censuses (1850–1950). *Journal de la Société des Américanistes* 95(2): 207–34.

Lubinsky, Mark S. 1993. Degenerate heredity: the history of a doctrine in medicine and biology. *Perspectives in Biology and Medicine* 37(1): 74–90.

Lynn, Richard. 2006. *Race differences in intelligence: an evolutionary analysis*. Augusta, GA: Washington Summit Publishers.

M'charek, Amade. 2005. *The Human Genome Diversity Project: an ethnography of scientific practice*. Cambridge: Cambridge University Press.

———. 2008. Silent witness, articulate collective: DNA evidence and the inference of visible traits. *Bioethics* 22(9): 519–28.

———. 2013. Beyond fact or fiction: on the materiality of race in practice. *Cultural Anthropology* 28(3): 420–42.

M'charek, Amade, Katharina Schramm and David Skinner. 2014a. Technologies of belonging: the absent presence of race in Europe. *Science, Technology & Human Values* 39(4): 459–67.

———. 2014b. Topologies of race: doing territory, population and identity in Europe. *Science, Technology & Human Values* 39(4): 468–87.

MacCallum, Mungo. 2002. *Girt by sea: Australia, the refugees and the politics of fear*. Melbourne: Black Incorporated.

Macey, Marie. 1995. 'Same race' adoption policy: anti-racism or racism? *Journal of Social Policy* 24(4): 473–91.

Macpherson, William. 1999. *The Stephen Lawrence inquiry*. Cm 4262-I. London: The Stationery Office.

Magubane, Zine. 2001. Which bodies matter? Feminism, poststructuralism, race, and the curious theoretical odyssey of the 'Hottentot Venus'. *Gender and Society* 15(6): 816–34.

Malcolm, Dominic. 2012. *Sport and sociology*. London: Routledge.

Malik, Kenan. 2009. *Strange fruit: why both sides are wrong in the race debate*. London: Oneworld.

Mangcu, Xolela. 2003. The state of race relations in post-apartheid South Africa. In *State of the nation: South Africa 2003–2004*, edited by John Daniel, Adam Habib and Roger Southall, 105–17. Cape Town: HSRC Press.

Marks, Jonathan. 1995. *Human biodiversity: genes, race, and history*. New York: Aldine de Gruyter.

  2001. 'We're going to tell these people who they really are': science and relatedness. In *Relative values: reconfiguring kinship studies*, edited by Sarah Franklin and Susan McKinnon, 355–83. Durham: Duke University Press.

  2003. *What it means to be 98% chimpanzee: apes, people, and their genes*. Berkeley: University of California Press.

Marre, Diana. 2007. 'I want her to learn her language and maintain her culture': transnational adoptive families' views of 'cultural origins'. In *Race, ethnicity and nation: perspectives from kinship and genetics*, edited by Peter Wade, 73–93. Oxford: Berghahn.

Martínez, María Elena. 2008. *Genealogical fictions: limpieza de sangre, religion, and gender in colonial Mexico*. Stanford: Stanford University Press.

Martínez-Echazábal, Lourdes. 1998. Mestizaje and the discourse of national/cultural identity in Latin America, 1845–1959. *Latin American Perspectives* 25(3): 21–42.

Marx, Anthony. 1998. *Making race and nation: a comparison of South Africa, the United States, and Brazil*. Cambridge: Cambridge University Press.

Massey, Douglas S., and Nancy A. Denton. 1993. *American apartheid: segregation and the making of the underclass*. Cambridge, MA: Harvard University Press.

Matory, J. Lorand. 1999. The English professors of Brazil: on the diasporic roots of the Yorùbá nation. *Comparative Studies in Society and History* 41(1): 72–103.

Maxwell, Rahsaan. 2012. *Ethnic minority migrants in Britain and France: integration trade-offs*. Cambridge: Cambridge University Press.

Mayr, Ernst. 1982. *The growth of biological thought: diversity, evolution and inheritance*. Cambridge, MA: Belknap Press of Harvard University Press.

McCallum, Cecilia. 2001. *Gender and sociality in Amazonia: how real people are made*. Oxford: Berg.

McClintock, Anne. 1995. *Imperial leather: race, gender and sexuality in the colonial contest*. London: Routledge.

McLoughlin, Seán. 2006. Writing a BrAsian city. In *A postcolonial people: South Asians in Britain*, edited by Nasreen Ali, Virinder S. Kalra and Salman Sayyid, 110–40. London: Hurst & Company.

  2014. Discrepant representations of multi-Asian Leicester: institutional discourse and every-day life in the model multicultural city. In *Writing the city in British-Asian diasporas*, edited by Seán McLoughlin, William Gould, Ananya Jahanara Kabir and Emma Tomalin, 89–113. London: Routledge.

McLoughlin, Seán, William Gould, Ananya Jahanara Kabir et al., eds. 2014. *Writing the city in British-Asian diasporas*. London: Routledge.

Mehta, Uday Singh. 1997. Liberal strategies of exclusion. In *Tensions of empire: colonial cultures in a bourgeois world*, edited by Frederick Cooper and Ann L. Stoler, 59–86. Berkeley: University of California Press.

Menchaca, Martha. 2007. Latinas/os and the mestizo racial heritage of Mexican Americans. In *A companion to Latina/o Studies*, edited by Juan Flores and Renato Rosaldo, 313–24. Oxford: Blackwell.

Merchant, Carolyn. 2006. The scientific revolution and the death of nature. *Isis* 97(3): 513–33.

Mignolo, Walter. 2005. *The idea of Latin America*. Oxford: Blackwell.

2011. *The darker side of Western modernity: global futures, decolonial options*. Durham: Duke University Press.

Migration Policy Centre. 2013. *MPC migration profile – Russia*. Florence: Migration Policy Centre, European University Institute.

Migration Policy Institute. 2014. *U.S. immigration trends*. Migration Policy Institute [cited 19 February 2014]. Available from http://www.migrationpolicy.org/programs/data-hub/us-immigration-trends#source.

Miles, Robert. 1989. *Racism*. London: Routledge.

Miles, Robert, and Malcolm Brown. 2003. *Racism*. 2nd edn. London: Routledge.

Mill, John Stuart. 1859. *On liberty*. London: Walter Scott Publishing Co.

Ministerio de Sanidad y Consumo. 1996. Real Decreto 412/1996, de 1 de marzo. *Boletín Oficial del Estado* 72: 11253–6.

Ministry of Justice. 2011. *Statistics on race and the criminal justice system 2010*. London: Ministry of Justice.

Mintz, Sidney. 1985. *Sweetness and power: the place of sugar in modern history*. New York: Sifton.

Modood, Tariq. 2007. *Multiculturalism: a civic idea*. Cambridge: Polity.

Modood, Tariq, and Stephen May. 2001. Multiculturalism and education in Britain: an internally contested debate. *International Journal of Educational Research* 35(3): 305–17.

Modood, Tariq, and Pnina Werbner, eds. 1997. *The politics of multiculturalism in the new Europe: racism, identity and community*. London: Zed.

Moleke, Percy. 2003. The state of the labour market in contemporary South Africa. In *State of the nation: South Africa 2003–2004*, edited by John Daniel, Adam Habib and Roger Southall, 204–24. Cape Town: HSRC Press.

Montagu, Ashley. 1942. *Man's most dangerous myth: the fallacy of race*. New York: Columbia University Press.

Montagu, Ashley, ed. 1964. *The concept of race*. New York: Free Press.

Moreno Figueroa, Mónica. 2008. Historically-rooted transnationalism: slightedness and the experience of racism in Mexican families. *Journal of Intercultural Studies* 29(3): 283–97.

2010. Distributed intensities: whiteness, mestizaje and the logics of Mexican racism. *Ethnicities* 10(3): 387–401.

2012. 'Linda morenita': skin colour, beauty and the politics of mestizaje in Mexico. In *Cultures of colour: visual, material, textual*, edited by Chris Horrocks, 167–80. Oxford: Berghahn.

Moreton-Robinson, Aileen, Maryrose Casey and Fiona Nicoll. 2008. *Transnational whiteness matters*. Lanham, MD: Lexington.

Morgan, Philip D. 1998. *Slave counterpoint: black culture in the eighteenth-century Chesapeake and Lowcountry*. Chapel Hill: University of North Carolina Press.

Morning, Ann. 2008. Ethnic classification in global perspective: a cross-national survey of the 2000 census round. *Population Research and Policy Review* 27(2): 239–72.

Mosse, George L. 1985. *Toward the final solution: a history of European racism*. Madison: University of Wisconsin Press.

Moynihan, Daniel P. 1965. *The Negro family: the case for national action*. Washington, DC: Office of Policy Planning and Research, US Department of Labor.

Mukhopadhyay, Carol C., and Yolanda T. Moses. 1997. Reestablishing 'race' in anthropological discourse. *American Anthropologist* 99(3): 517–33.

Nagel, Joane. 2003. *Race, ethnicity, and sexuality: intimate intersections, forbidden frontiers*. Oxford: Oxford University Press.

Nahman, Michal. 2006. Materializing Israeliness: difference and mixture in transnational ova donation. *Science as Culture* 15(3): 199–213.

2010. 'Embryos are our baby': abridging hope, body and nation in transnational ova donation In *Technologized images, technologized bodies*, edited by Jeanette Edwards, Penny Harvey and Peter Wade, 185–210. Oxford: Berghahn.

Nash, Catherine. 2013. Genome geographies: mapping national ancestry and diversity in human population genetics. *Transactions of the Institute of British Geographers* 38(2): 193–206.

Nash, Gary. 1995. The hidden history of mestizo America. *Journal of American History* 82(3): 941–64.

National Urban League. 2014. *One nation underemployed: jobs rebuild America. 2014 state of Black America*. New York: National Urban League.

Nederveen Pieterse, Jan. 1992. *White on black: images of Africa and blacks in Western popular culture*. New Haven: Yale University Press.

Nelson, Alondra. 2008. The factness of diaspora: the social sources of genetic genealogy. In *Revisiting race in a genomic age*, edited by Barbara A. Koenig, Sandra Soo-Jin Lee and Sarah S. Richardson, 253–68. New Brunswick: Rutgers University Press.

Nelson, Diane M. 1999. *A finger in the wound: body politics in quincentennial Guatemala*. Berkeley: California University Press.

Ng'weno, Bettina. 2007. *Turf wars: territory and citizenship in the contemporary state*. Stanford: Stanford University Press.

Niehaus, Isak. 2013. Anthropology and whites in South Africa: response to an unreasonable critique. *Africa Spectrum* 48(1): 117–27.

Nirenberg, David. 2009. Was there race before modernity? The example of 'Jewish' blood in late medieval Spain. In *The origins of racism in the West*, edited by Miriam Eliav-Feldon, Benjamin Isaac and Joseph Ziegler, 232–64. Cambridge: Cambridge University Press.

Nobles, Melissa. 2000. *Shades of citizenship: race and the census in modern politics*. Stanford: Stanford University Press.

Nogueira, Oracy. 2008 [1959]. Skin color and social class. *Vibrant* 5(1): xxix–li.

Nyamnjohm, Francis B. 2012. Blinded by sight: divining the future of anthropology in Africa. *Africa Spectrum* 47(2–3): 63–92.

Observatoire des inégalités. 2011. *Le taux de chômage selon la nationalité* [cited 29 May 2014]. Available from http://www.inegalites.fr/spip.php?article86&id_groupe=17&id_mot=112&id_rubrique=97.

n.d. *Racial discrimination less present at job interview stage?* [cited 29 May 2014]. Available from http://www.discriminations.inegalites.fr/spip.php?article136&lang=en.

OECD. n.d. *Country-of-birth database* [cited 18 February 2014]. Available from www.oecd.org/migration/mig/34792376.xls.

Office for National Statistics. 2013. *Non-UK born census populations* 1951–2011 [cited 18 February 2014]. Available from http://www.ons.gov.uk/ons/rel/census/2011-census-analysis/immigration-patterns-and-characteristics-of-non-uk-born-population-groups-in-england-and-wales/non-uk-born-census-populations-1951–2011–full-infographic.html.

n.d. *Neighbourhood statistics* [cited 6 June 2014]. Available from http://neighbourhood.statistics.gov.uk/.

Ofsted. 2014. *School report: Park View School Academy of Mathematics and Science*. Manchester: Ofsted.

Olumide, Jill. 2002. *Raiding the gene pool: the social construction of mixed race*. London: Pluto Press.

Omi, Michael, and Howard Winant. 1986. *Racial formation in the United States: from the 1960s to the 1980s*. New York: Routledge.

1994. *Racial formation in the United States: from the 1960s to the 1990s.* 2nd edn. New York: Routledge.

Ong, Aihwa, and Stephen J. Collier, eds. 2008. *Global assemblages: technology, politics, and ethics as anthropological problems.* Oxford: Blackwell.

Open Society Justice Initiative. 2006. *Ethnic profiling in the Moscow Metro.* New York: Open Society Institute, JURIX.

2009. *Profiling minorities: a study of stop-and-search practices in Paris.* New York: Open Society Institute.

Ossorio, Pilar, and Troy Duster. 2005. Race and genetics: controversies in biomedical, behavioral, and forensic sciences. *American Psychologist* 60(1): 115–28.

Overing, Joanna. 1993. Death and the loss of civilized predation among the Piaroa of the Orinoco Basin. *L'Homme* 33(126–8): 191–211.

1996. Who is the mightiest of them all? Jaguar and conquistador in Piaroa images of alterity and identity. In *Monsters, tricksters and sacred cows: animal tales and American identities,* edited by A. James Arnold, 50–79. Charlottesville: University of Virginia Press.

Owen, Charlie. 2001. 'Mixed race' in official statistics. In *Rethinking 'mixed race',* edited by David Parker and Miri Song, 134–53. London: Pluto Press.

Pagden, Anthony. 1982. *The fall of natural man: the American Indian and the origins of comparative ethnology.* Cambridge: Cambridge University Press.

Pálsson, Gísli. 2007. *Anthropology and the new genetics.* Cambridge: Cambridge University Press.

Pan Ké Shon, Jean-Louis. 2011. La ségrégation des immigrés en France: état des lieux. *Population et Sociétés* 477: 1–4.

Parker, David, and Miri Song, eds. 2001. *Rethinking 'mixed race'.* London: Pluto Press.

Paschel, Tianna. 2013. 'The Beautiful Faces of my Black People': race, ethnicity and the politics of Colombia's 2005 census. *Ethnic and Racial Studies* 36(10): 1544–63.

forthcoming. *Becoming black political subjects: movements, global alignments and black rights in Colombia and Brazil.*

Passel, Jeffrey S., and D'Vera Cohn. 2009. *A portrait of unauthorized immigrants in the United States.* Washington, DC: Pew Hispanic Center.

Pattillo, Mary. 2007. *Black on the block: the politics of race and class in the city.* Chicago: University of Chicago Press.

2009. Revisiting Loïc Wacquant's Urban Outcasts. *International Journal of Urban and Regional Research* 33(3): 858–64.

Pattillo-McCoy, Mary. 2000. *Black picket fences: privilege and peril among the black middle class.* Chicago: University of Chicago Press.

Paul, Kathleen. 1997. *Whitewashing Britain: race and citizenship in the postwar era.* Ithaca: Cornell University Press.

Peach, Ceri. 2009. Slippery segregation: discovering or manufacturing ghettos? *Journal of Ethnic and Migration Studies* 35(9): 1381–95.

Pena, Sérgio D. J., Giuliano Di Pietro, Mateus Fuchshuber-Moraes et al. 2011. The genomic ancestry of individuals from different geographical regions of Brazil is more uniform than expected. *PLoS ONE* 6(2): e17063.

Pierre, Jemima. 2013. *The predicament of blackness: postcolonial Ghana and the politics of race.* Chicago: University of Chicago Press.

Pinho, Patricia de Santana. 2010. *Mama Africa: reinventing blackness in Bahia.* Durham: Duke University Press.

Pitarch, Pedro, Shannon Speed and Xochitl Leyva-Solano, eds. 2008. *Human rights in the Maya region: global politics, cultural contentions, and moral engagements.* Durham: Duke University Press.

Poole, Deborah. 1997. *Vision, race and modernity: a visual economy of the Andean image world*. Princeton: Princeton University Press.

Poole, Stafford. 1999. The politics of limpieza de sangre: Juan de Ovando and his circle in the reign of Philip II. *Americas* 55(3): 359–89.

Pooley, Rachel. 2010. *Constructing race with images* [cited 12 June 2014]. Available from http://race-in-colonial-mexico.net/colonialrace/.

Poovey, Mary. 1998. *A history of the modern fact: problems of knowledge in the sciences of wealth and society*. Chicago: University of Chicago Press.

Population Studies Center. 2010. *New racial segregation measures for large metropolitan areas: analysis of the 1990–2010 decennial censuses*. University of Michigan [cited 9 April 2014]. Available from http://www.psc.isr.umich.edu/dis/census/segregation2010.html.

Porqueres i Gené, Enric. 2007. Kinship language and the dynamics of race: the Basque case. In *Race, ethnicity and nation: perspectives from kinship and genetics*, edited by Peter Wade, 125–44. Oxford: Berghahn.

Povinelli, Elizabeth A. 2002. *The cunning of recognition: indigenous alterities and the making of Australian multiculturalism*. Durham: Duke University Press.

Powdermaker, Hortense. 1939. *After freedom: a cultural study in the Deep South*. New York: Viking.

Powell, Enoch. 1969. *Freedom and reality*. Kingswood: Elliot Right Way Books.

Queen's University. 2014. *Multiculturalism policy index* [cited 21 February 2014]. Available from http://www.queensu.ca/mcp/.

Qureshi, Sadiah. 2004. Displaying Sara Baartman, the 'Hottentot Venus'. *History of Science* 42: 233–57.

Radcliffe, Sarah. 1990. Ethnicity, patriarchy and incorporation into the nation: female migrants as domestic servants in southern Peru. *Environment and Planning D: Society and Space* 8(4): 379–93.

Radin, Paul. 1929. History of ethnological theories. *American Anthropologist* 31: 9–33.

Ragoné, Helena. 1998. Incontestable motivations. In *Reproducing reproduction: kinship, power, and technological innovation*, edited by Sarah Franklin and Helena Ragoné, 118–31. Philadelphia: University of Pennsylvania Press.

Rahier, Jean Muteba. 2003. Introduction: mestizaje, mulataje, mestiçagem in Latin American ideologies of national identities. *Journal of Latin American Anthropology* 8(1): 40–50.

Ramos-Zayas, Ana Y. 2003. *National performances: the politics of class, race and space in Puerto Rican Chicago*. Chicago: University of Chicago Press.

Randerson, James. 2006. DNA of 37% of black men held by police. *The Guardian*, 5 January.

Rappaport, Joanne. 2014. *The disappearing mestizo: configuring difference in the colonial New Kingdom of Granada*. Durham: Duke University Press.

Reardon, Jenny. 2005. *Race to the finish: identity and governance in an age of genomics*. Princeton: Princeton University Press.

———. 2008. Race without salvation: beyond the science/society divide in genomic studies of human diversity. In *Revisiting race in a genomic age*, edited by Barbara A. Koenig, Sandra Soo-Jin Lee and Sarah S. Richardson, 304–19. New Brunswick: Rutgers University Press.

Reeves, Frank. 1983. *British racial discourse: a study of British political discourse about race and race-related matters*. Cambridge: Cambridge University Press.

Reeves, Madeleine. 2013. Kak stanoviatsia 'chernym' v Moskve: praktiki vlasti i sushchest-vovanie migrantov v teni zakona [Becoming 'black' in Moscow: documentary regimes and migrant life in the shadow of law]. In *Grazhdanstvo i immigratsiia: kontseputal'noe, istoriches-koe i institutsional'noe izmenenie* [Citizenship and immigration in Russia: conceptual, historical and institutional dimensions], edited by Vladmir Malakhov, 146–77. Moscow: Russian Academy of Sciences/Kanon+.

Reichmann, Rebecca, ed. 1999. *Race in contemporary Brazil: from indifference to inequality.* University Park: Pennsylvania State University Press.

Reiter, Bernd, and Gladys L. Mitchell, eds. 2010. *Brazil's new racial politics.* Boulder: Lynne Rienner.

Reuter, Edward. 1918. *The mulatto in the United States, including a study of the role of mixed-blood races throughout the world.* Boston: Richard G. Badger.

Ribeiro Corossacz, Valeria. 2015. Whiteness, maleness and power: a study in Rio de Janeiro. *Latin American and Caribbean Ethnic Studies* 10(2).

Ritvo, Harriet. 1997. *The platypus and the mermaid and other figments of the classifying imagination.* Cambridge, MA: Harvard University Press.

Robb, Peter. 1995a. Introduction: South Asia and the concept of race. In *The concept of race in South Asia,* edited by Peter Robb, 1–76. Oxford: Oxford University Press.

Robb, Peter, ed. 1995b. *The concept of race in South Asia.* Oxford: Oxford University Press.

Roberts, Elizabeth F. S. 2012. *God's laboratory: assisted reproduction in the Andes.* Berkeley: University of California Press.

2013. Assisted existence: an ethnography of being in Ecuador. *Journal of the Royal Anthropological Institute* 19(3): 562–80.

Rodríguez, Clara E. 1994. Challenging racial hegemony: Puerto Ricans in the United States. In *Race,* edited by Steven Gregory and Roger Sanjek, 131–45. New Brunswick: Rutgers University Press.

2000. *Changing race: Latinos, the census, and the history of ethnicity in the United States.* New York: New York University Press.

Rodriguez, Richard. 2002. *Brown: the last discovery of America.* New York: Viking.

Rodríguez Garavito, César, Tatiana Alfonso Sierra and Isabel Cavelier Adarve. 2009. *Raza y derechos humanos en Colombia: informe sobre discriminación racial y derechos de la población afrocolombiana.* Bogotá: Universidad de los Andes, Facultad de Derecho, Centro de Investigaciones Sociojurídicas (CIJUS), Observatorio de Discriminación Racial, Ediciones Uniandes.

Roediger, David R. 2006. *Working toward whiteness: how America's immigrants became white.* New York: Basic Books.

Roitman, Karem. 2009. *Race, ethnicity and power in Ecuador: the manipulation of mestizaje.* Boulder: First Forum Press.

Rosenberg, Noah A., Jonathan K. Pritchard, James L. Weber et al. 2002. Genetic structure of human populations. *Science* 298: 2381–5.

Safa, Helen I. 2005. Challenging *mestizaje*: a gender perspective on indigenous and Afrodescendant movements in Latin America. *Critique of Anthropology* 25(3): 307–30.

Safi, Mirna. 2009. La dimension spatiale de l'intégration: évolution de la ségrégation des populations immigrées en France entre 1968 et 1999. *Revue française de sociologie* 50: 521–52.

Saldaña-Tejeda, Abril. 2012. 'Why should I not take an apple or a fruit if I wash their underwear?' Food, social classification and paid domestic work in Mexico. *Journal of Intercultural Studies* 33(2): 121–37.

Sánchez Taylor, Jacqueline. 2010. Sex tourism and inequalities. In *Tourism and inequality: problems and prospects,* edited by Stroma Cole and Nigel Morgan, 49–66. Wallingford: CABI.

Sanford, Victoria. 2003. *Buried secrets: truth and human rights in Guatemala.* New York: Palgrave Macmillan.

Sanjek, Roger. 1998. *The future of us all: race and neighborhood politics in New York City.* Ithaca: Cornell University Press.

Sankar, Pamela. 2012. Forensic DNA phenotyping: continuity and change in the history of race, genetics, and policing. In *Genetics and the unsettled past: the collision of DNA, race, and history,* edited by Keith Wailoo, Alondra Nelson and Catherine Lee, 104–13. New Brunswick: Rutgers University Press.

Sansone, Livio. 2003. *Blackness without ethnicity: constructing race in Brazil.* Basingstoke: Palgrave Macmillan.

Santos, Ricardo Ventura, Peter H. Fry, Simone Monteiro et al. 2009. Color, race and genomic ancestry in Brazil: dialogues between anthropology and genetics. *Current Anthropology* 50(6): 787–819.

Santos-Granero, Fernando. 2009. Amerindian constructional views of the world. In *The occult life of things: native Amazonian theories of materiality and personhood*, edited by Fernando Santos-Granero, 1–29. Tucson: University of Arizona Press.

2012. Beinghood and people-making in native Amazonia: a constructional approach with a perspectival coda. *HAU: Journal of Ethnographic Theory* 2(1): 181–211. Accessed 1 October 2013. Available from http://www.haujournal.org/index.php/hau/article/view/25/108.

Saunders, A. C. de C. M. 1982. *A social history of black slaves and freedmen in Portugal, 1441–1555.* Cambridge: Cambridge University Press.

Schaffer, Gavin. 2008. *Racial science and British society, 1930–62.* Basingstoke: Palgrave Macmillan.

Schain, Martin A. 2012. *The politics of immigration in France, Britain and the United States: a comparative study*. 2nd edn. New York: Palgrave Macmillan.

Scheper-Hughes, Nancy. 2000. The global traffic in human organs. *Current Anthropology* 41(2): 191–224.

Schramm, Katharina. 2012. Genomics en route: ancestry, heritage, and the politics of identity across the Black Atlantic. In *Identity politics and the new genetics: re/creating categories of difference and belonging*, edited by Katharina Schramm, David Skinner and Richard Rottenburg, 167–92. Oxford: Berghahn.

Schwartz-Marín, Ernesto. n.d. Explaining the visible and the invisible: lay knowledge of genetics, ancestry, physical appearance and race in Colombia. Unpublished work. Durham: Durham University.

Schwartzman, Luisa Farah. 2007. Does money whiten? Intergenerational changes in racial classification in Brazil. *American Sociological Review* 72: 940–63.

2009. Seeing like citizens: unofficial understandings of official racial categories in a Brazilian university. *Journal of Latin American Studies* 41(2): 221–50.

Scott, James. 1985. *Weapons of the weak: everyday forms of peasant resistance.* New Haven: Yale University Press.

Seekings, Jeremy. 2010. *Race, class and inequality in the South African city.* CSSR Working Paper Series 283. Cape Town: Centre for Social Science Research, University of Cape Town.

Seekings, Jeremy, and Nicoli Nattrass. 2008. *Class, race, and inequality in South Africa.* New Haven: Yale University Press.

Seitles, Marc. 1996. The perpetuation of residential racial segregation in America: historical discrimination, modern forms of exclusion, and inclusionary remedies. *Journal of Land Use & Environmental Law* 14(1): 89–123.

Selka, Stephen. 2007. *Religion and the politics of ethnic identity in Bahia, Brazil.* Gainesville: University Press of Florida.

Selman, Peter. 2002. Intercountry adoption in the new millennium: the 'quiet migration' revisited. *Population Research and Policy Review* 21: 205–25.

2013. *Statistics based on data provided by 23 receiving states, compiled by Professor Selman (7 October 2013).* The Hague: Hague Conference on Private International Law.

Senghor, Léopold. 2003. The contribution of the black man. In *Race and racism in continental philosophy*, edited by Robert Bernasconi and Sybol Cook, 287–303. Bloomington: Indiana University Press.

Serre, David, and Svante Pääbo. 2004. Evidence for gradients of human genetic diversity within and among continents. *Genome Research* 14(9): 1679–85.

Shanklin, Eugenia. 1994. *Anthropology and race.* Belmont, CA: Wadsworth Publishing.

Sharp, John. 2001. The question of cultural difference: anthropological perspectives in South Africa. *South African Journal of Ethnology* 24(3): 67–74.

Sharp, John, and Rehana Vally. 2009. Unequal 'cultures'? Racial integration at a South African university. *Anthropology Today* 25(3): 3–6.

Sheriff, Robin E. 2001. *Dreaming equality: color, race, and racism in urban Brazil*. New Brunswick: Rutgers University Press.

　2003. Embracing race: deconstructing mestiçagem in Rio de Janeiro. *Journal of Latin American Anthropology* 8(1): 86–115.

Sieder, Rachel, ed. 2002. *Multiculturalism in Latin America: indigenous rights, diversity and democracy*. Basingstoke: Palgrave Macmillan.

Silva-Zolezzi, Irma, Alfredo Hidalgo-Miranda, Jesus Estrada-Gil et al. 2009. Analysis of genomic diversity in Mexican Mestizo populations to develop genomic medicine in Mexico. *Proceedings of the National Academy of Sciences* 106(21): 8611–16.

Silverblatt, Irene. 1994. Andean witches and virgins: seventeenth-century nativism and subversive gender ideologies. In *Women, 'race,' and writing in the early modern period*, edited by Margo Hendricks and Patricia Parker, 259–71. London: Routledge.

　2004. *Modern Inquisitions: Peru and the colonial origins of the civilized world*. Durham: Duke University Press.

Silverstein, Paul A. 2004. *Algeria in France: transpolitics, race, and nation*. Bloomington: Indiana University Press.

　2005. Immigrant racialization and the new savage slot: race, migration, and immigration in the New Europe. *Annual Review of Anthropology* 34(1): 363–84.

Simpson, Ludi. 2012. *More segregation or more mixing?* Manchester: Centre on Dynamics of Ethnicity, University of Manchester.

Sio, Arnold A. 1976. Race, colour, and miscegenation: the free coloured of Jamaica and Barbados. *Caribbean Studies* 16(1): 5–21.

Skidmore, Thomas. 1990. Racial ideas and social policy in Brazil, 1870–1940. In *The idea of race in Latin America, 1870–1940*, edited by Richard Graham, 7–36. Austin: University of Texas Press.

　2003. Racial mixture and affirmative action: the cases of Brazil and the United States. *American Historical Review* 108(5): 1391–6.

Slocum, Rachel B., and Arun Saldanha, eds. 2013. *Geographies of race and food: fields, bodies, markets*. Farnham: Ashgate.

Smart, Andrew. 2005. Practical concerns that arise from using race/ethnicity as 'the most reliable proxy available'. *Critical Public Health* 15(1): 75–6.

Smedley, Audrey. 1993. *Race in North America: origin and evolution of a worldview*. Boulder and Oxford: Westview Press.

　1998. 'Race' and the construction of human identity. *American Anthropologist* 100(3): 690–702.

Smith, Katherine. 2012. *Fairness, class and belonging in contemporary England*. Basingstoke: Palgrave Macmillan.

Snowden, Frank M. 1983. *Before colour prejudice: the ancient view of blacks*. Cambridge, MA: Harvard University Press.

Sollors, Werner. 1981. Theory of American ethnicity, or: '? S Ethnic?/Ti and American/Ti, De or United (W) States S S1 and Theor?'. *American Quarterly* 33(3): 257–83.

Soper, Kate. 1995. *What is nature? Culture, politics and the non-human*. Oxford: Blackwell.

Stack, Carol. 1974. *All our kin: strategies for survival in a black community*. New York: Harper & Row.

Statistics South Africa. 2004. *Census 2001: concepts and definitions*. Pretoria: Statistics South Africa.

　Census 2011: census in brief. Pretoria: Statistics South Africa.

Stepan, Nancy. 1982. *The idea of race in science: Great Britain, 1800–1960*. London: Macmillan in association with St Antony's College, Oxford.

Stepan, Nancy Leys. 1991. *'The hour of eugenics': race, gender and nation in Latin America*. Ithaca: Cornell University Press.

Stephen, Lynn. 2007. *Transborder lives: indigenous Oaxacans in Mexico, California, and Oregon.* Durham: Duke University Press.

Stern, Alexandra Minna. 2009. Eugenics and racial classification in modern Mexican America. In *Race and classification: the case of Mexican America*, edited by Ilona Katzew and Susan Deans-Smith, 151–73. Stanford: Stanford University Press.

Stocking, George. 1982. *Race, culture and evolution: essays on the history of anthropology.* 2nd edn. Chicago: University of Chicago Press.

Stolcke, Verena. 1995. Talking culture: new boundaries, new rhetorics of exclusion in Europe. *Current Anthropology* 36(1): 1–23.

Stoler, Ann Laura. 1995. *Race and the education of desire: Foucault's History of Sexuality and the colonial order of things.* Durham: Duke University Press.

  2002. *Carnal knowledge and imperial power: race and the intimate in colonial rule.* Berkeley: University of California Press.

Strathern, Marilyn. 1980. No nature, no culture: the Hagen case. In *Nature, culture, gender*, edited by Carol P. MacCormack and Marilyn Strathern, 174–222. Cambridge: Cambridge University Press.

  1992. *After nature: English kinship in the late twentieth century.* Cambridge: Cambridge University Press.

Streicker, Joel. 1995. Policing boundaries: race, class, and gender in Cartagena, Colombia. *American Ethnologist* 22(1): 54–74.

Štrkalj, Goran. 2007. The status of the race concept in contemporary biological anthropology: a review. *The Anthropologist* 9(1): 73–8.

Stubbe, Hans. 1972. *History of genetics: from prehistoric times to the rediscovery of Mendel's laws.* Translated by T. R. W. Waters. Cambridge, MA: MIT Press.

Sue, Christina A. 2013. *Land of the cosmic race: race mixture, racism, and blackness in Mexico.* New York: Oxford University Press.

Taguieff, Pierre-André. 1990. The new cultural racism in France. *Telos* 83: 109–22.

Takezawa, Yasuko. 2005. Transcending the Western paradigm of the idea of race. *Japanese Journal of American Studies* 16: 5–30.

Taussig, Karen-Sue, and Sahra Elizabeth Gibbon. 2013. Introduction: Public health genomics – anthropological interventions in the quest for molecular medicine. *Medical Anthropology Quarterly* 27(4): 471–88.

Tavan, Gwenda. 2005. *The long, slow death of white Australia.* Melbourne: Scribe Publications.

Taylor, Charles, and Amy Gutmann. 1994. *Multiculturalism and 'the politics of recognition'.* Princeton: Princeton University Press.

Taylor, Chloë. 2011. Race and racism in Foucault's Collège de France lectures. *Philosophy Compass* 6(11): 746–56.

Telles, Edward E. 2004. *Race in another America: the significance of skin color in Brazil.* Princeton: Princeton University Press.

Telles, Edward E., and René Flores. 2013. Not just color: whiteness, nation and status in Latin America. *Hispanic American Historical Review* 93(3): 411–49.

Thomas, Deborah. 2004. *Modern blackness: nationalism, globalisation and the politics of culture in Jamaica.* Durham: Duke University Press.

Thomas, James M. 2010. The racial formation of medieval Jews: a challenge to the field. *Ethnic and Racial Studies* 33(10): 1737–55.

Thompson, Lloyd A. 1989. *Romans and blacks.* London: Routledge.

Thomson, Sinclair. 2011. Was there race in colonial Latin America? Identifying selves and others in the insurgent Andes. In *Histories of race and racism: the Andes and Mesoamerica from colonial times to the present*, edited by Laura Gotkowitz, 72–91. Durham: Duke University Press.

Todorov, Tzvetan. 1993. *On human diversity: nationalism, racism and exoticism in French thought*. Cambridge, MA: Harvard University Press.

Tolnay, Stuart E., and E. M. Beck. 1995. *A festival of violence: an analysis of Southern lynchings, 1882–1930*. Urbana: University of Illinois Press.

Tomich, Dale W. 2004. *Through the prism of slavery: labor, capital, and world economy*. Lanham, MD: Rowman & Littlefield.

Trimier, Jacqueline. 2003. The myth of authenticity: personhood, traditional culture and African philosophy. In *From Africa to Zen: an invitation to world philosophy*, edited by Robert C. Solomon and Kathleen M. Higgins, 173–200. Lanham, MD: Rowman & Littlefield.

Trojan Horse Review Group. 2014. *Report to Leader of Birmingham City Council*. Birmingham: Birmingham City Council.

Twinam, Ann. 1999. *Public lives, private secrets: gender, honor, sexuality and illegitimacy in colonial Spanish America*. Stanford: Stanford University Press.

Twine, France Winddance. 1998. *Racism in a racial democracy: the maintenance of white supremacy in Brazil*. New Brunswick: Rutgers University Press.

——— 2000. Bearing blackness in Britain: the meaning of racial difference for white birth mothers of African-descent children. In *Ideologies and technologies of motherhood: race, class, sexuality, nationalism*, edited by Helena Ragoné and France Winddance Twine, 76–108. London: Routledge.

Tyler, Katharine. 2005. The genealogical imagination: the inheritance of interracial identities. *Sociological Review* 53(3): 476–94.

——— 2007. Race, genetics and inheritance: reflections upon the birth of 'black' twins to a 'white' IVF mother. In *Race, ethnicity and nation: perspectives from kinship and genetics*, edited by Peter Wade, 33–51. Oxford: Berghahn.

——— 2012. *Whiteness, class and the legacies of empire: on home ground*. Basingstoke: Palgrave Macmillan.

Tyler-Smith, Chris, and Yali Xue. 2012. A British approach to sampling. *European Journal of Human Genetics* 20(2): 129–30.

UNESCO. 1952. *The race concept: results of an inquiry*. Paris: UNESCO.

US Bureau of the Census. 1949. *Historical statistics of the United States, 1789–1945*. Washington, DC: US Bureau of the Census.

——— 1975. *Historical statistics of the United States: colonial times to 1970*. Washington, DC: US Bureau of the Census.

——— 2004. *Measuring America: the decennial censuses from 1790 to 2000*. Washington, DC: Government Printing Office.

——— 2013a. *Historical income tables: households* [cited 12 April 2014]. Available from http://www.census.gov/hhes/www/income/data/historical/household/.

——— 2013b. *Historical poverty tables* [cited 12 April 2014]. Available from http://www.census.gov/hhes/www/poverty/data/historical/index.html.

——— 2013c. *Race* [cited 11 June 2014]. Available from http://www.census.gov/topics/population/race.html.

——— n.d. *Statistical abstracts* [cited 7 March 2014]. Available from https://www.census.gov/prod/www/statistical_abstract.html.

US Department of State. 2013. *Trafficking in persons report 2013* [cited 10 June 2014]. Available from http://www.state.gov/j/tip/rls/tiprpt/2013/index.htm.

US Housing Scholars et al. 2008. *Racial segregation and housing discrimination in the United States*. Washington, DC: Poverty and Race Research Action Council and National Fair Housing Alliance.

van den Berghe, Pierre. 1979. *The ethnic phenomenon*. New York: Elsevier.

Vermeulen, Han F. 1995. Origins and institutionalization of ethnography and ethnology in Europe and the USA, 1771–1845. In *Fieldwork and footnotes: studies in the history of European anthropology*, edited by Han F. Vermeulen and Arturo Alvarez Roldán, 39–59. London: Routledge.

Viáfara López, Carlos Augusto. 2008. Diferencias raciales en el logro educativo y status ocupacional en el primer empleo, en la ciudad de Cali (Colombia). In *Pobreza, exclusión social y discriminación étnico-racial en América Latina y el Caribe*, edited by María del Carmen Zabala Argüelles, 85–120. Bogotá: Siglo del Hombre Editores, CLACSO.

Voegelin, Eric, and Klaus Vondung. 1998 [1933]. *The history of the race idea: from Ray to Carus*. Translated by Ruth Hein. Baton Rouge: Louisiana State University Press.

Volkman, Toby Alice. 2005a. Embodying Chinese culture: transnational adoption in North America. In *Cultures of transnational adoption*, edited by Toby Alice Volkman, 81–113. Durham: Duke University Press.

Volkman, Toby Alice, ed. 2005b. *Cultures of transnational adoption*. Durham: Duke University Press.

Wacquant, Loïc. 2008. *Urban outcasts: a comparative sociology of advanced marginality*. Cambridge: Polity Press.

Wade, Peter. 1993. *Blackness and race mixture: the dynamics of racial identity in Colombia*. Baltimore: Johns Hopkins University Press.

2000. *Music, race and nation: música tropical in Colombia*. Chicago: University of Chicago Press.

2002a. The Colombian Pacific in perspective. *Journal of Latin American Anthropology* 7(2): 2–33.

2002b. *Race, nature and culture: an anthropological perspective*. London: Pluto Press.

2004. Images of Latin American mestizaje and the politics of comparison. *Bulletin of Latin American Research* 23(1): 355–66.

2005. Rethinking mestizaje: ideology and lived experience. *Journal of Latin American Studies* 37: 1–19.

2009a. Defining blackness in Colombia. *Journal de la Société des Américanistes* 95(1): 165–84.

2009b. *Race and sex in Latin America*. London: Pluto Press.

2010. *Race and ethnicity in Latin America*. 2nd edn. London: Pluto Press.

2012. Race, kinship and the ambivalence of identity. In *Identity politics and the new genetics: re/creating categories of difference and belonging*, edited by Katharina Schramm, David Skinner and Richard Rottenburg, 79–96. Oxford: Berghahn.

2013a. Articulations of eroticism and race: domestic service in Latin America. *Feminist Theory* 14(2): 187–202.

2013b. Blackness, indigeneity, multiculturalism and genomics in Brazil, Colombia and Mexico. *Journal of Latin American Studies* 45(2): 205–33.

2013c. Brazil and Colombia: comparative race relations in South America. In *Racism and ethnic relations in the Portuguese-speaking world*, edited by Francisco Bethencourt and Adrian Pearce, 35–48. Oxford: Oxford University Press, British Academy.

Wade, Peter, Carlos López Beltrán, Eduardo Restrepo et al., eds. 2014. *Mestizo genomics: race mixture, nation, and science in Latin America*. Durham: Duke University Press.

Walker, Shaun. 2011. Russian immigration official sacked for promoting 'survival of white race'. *The Independent*, 22 April.

Wang, Qian, Goran Štrkalj and Li Sun. 2003. On the concept of race in Chinese biological anthropology: alive and well. *Current Anthropology* 44(3): 403.

Wang, Wendy. 2012. *The rise of intermarriage: rates, characteristics vary by race and gender*. Washington, DC: Pew Research Center.

Warner, W. Lloyd. 1936. American caste and class. *American Journal of Sociology* 42: 234–7.

Warren, Jonathan W. 2001. *Racial revolutions: antiracism and Indian resurgence in Brazil*. Durham: Duke University Press.

Warren, Kay B. 1998. *Indigenous movements and their critics: Pan-Maya activism in Guatemala*. Princeton: Princeton University Press.

Watson, James Lee, ed. 1977. *Between two cultures: migrants and minorities in Britain*. Oxford: Blackwell.

Weaver, Jace. 2014. *Red Atlantic: American indigenes and the making of the modern world, 1000–1927*. Chapel Hill: University of North Carolina Press.

Weiner, Michael. 1994. *Race and migration in imperial Japan*. London: Routledge.

1995. Discourses of race, nation and empire in pre-1945 Japan. *Ethnic and Racial Studies* 18(3): 433–56.

Weismantel, Mary. 2001. *Cholas and pishtacos: stories of race and sex in the Andes*. Chicago: University of Chicago Press.

Weismantel, Mary, and Stephen F. Eisenman. 1998. Race in the Andes: global movements and popular ontologies. *Bulletin of Latin American Research* 17(2): 121–42.

Werbner, Pnina. 2002. *Imagined diasporas among Manchester Muslims: the public performance of Pakistani transnational identity politics*. Oxford: James Currey.

Werbner, Pnina, and Muhammad Anwar, eds. 1991. *Black and ethnic leaderships in Britain: the cultural dimension of political action*. London: Routledge.

Whitten, Norman. 1986. *Black frontiersmen: a South American case*. 2nd edn. Prospect Heights, IL: Waveland Press.

Whitten, Norman, ed. 1981. *Cultural transformations and ethnicity in modern Ecuador*. Urbana: University of Illinois Press.

Wieviorka, Michel. 1995. *The arena of racism*. London: Sage.

Williams, Brackette. 1995. Classification systems revisited: kinship, caste, race and nationality as the flow of blood and the spread of rights. In *Naturalizing power: essays in feminist cultural analysis*, edited by Sylvia Yanagisako and Carol Delaney, 201–36. New York: Routledge.

Williams, Raymond. 1988. *Keywords: a vocabulary of culture and society*. London: Fontana.

Willsher, Kim. 2014. Le Pen's confidence vindicated. *Guardian Weekly*, 30 May.

Wilson, Peter J. 1973. *Crab antics: the social anthropology of English-speaking negro societies of the Caribbean*. New Haven: Yale University Press.

Wilson, Richard. 1995. *Maya resurgence in Guatemala: Q'echi' experiences*. Norman: University of Oklahoma Press.

Wilson, Richard A. 2001. *The politics of truth and reconciliation in South Africa: legitimizing the post-apartheid state*. Cambridge: Cambridge University Press.

Winant, Howard. 2002. *The world is a ghetto: race and democracy since World War II*. New York: Basic Books.

Winney, Bruce, Abdelhamid Boumertit, Tammy Day et al. 2012. People of the British Isles: preliminary analysis of genotypes and surnames in a UK-control population. *European Journal of Human Genetics* 20(2): 203–10.

Wise, M. Norton, ed. 1997. *The values of precision*. Princeton: Princeton University Press.

Wood, Martin, Jon Hales, Susan Purdon et al. 2009. *A test for racial discrimination in recruitment practice in British cities*. London: Department for Work and Pensions.

Work, Monroe N. 1931. *Negro yearbook: an annual encyclopedia of the Negro, 1931–1932*. Tuskegee Institute, AL: Negro Yearbook Publishing Company.

Wright, Winthrop. 1990. *Café con leche: race, class and national image in Venezuela*. Austin: University of Texas Press.

Yanagisako, Sylvia, and Carol Delaney, eds. 1995. *Naturalizing power: essays in feminist cultural analysis*. London: Routledge.

Yelvington, Kevin, ed. 2006. *Afro-Atlantic dialogues: anthropology in the diaspora*. Santa Fe, NM: School of American Research Press.

Yuval-Davis, Nira. 1997. *Gender and nation*. Thousand Oaks, CA: Sage.

Zakharov, Nikolay. 2013. Attaining whiteness: a sociological study of race and racialization in Russia. Ph.D. thesis, Sociology, Uppsala University, Uppsala.

INDEX